QUALITY CONTROL IN REMEDIAL SITE INVESTIGATION: HAZARDOUS AND INDUSTRIAL SOLID WASTE TESTING
Fifth Volume

A symposium sponsored by ASTM
Committee D-34 on Waste Disposal
New Orleans, Louisiana, 8–9 May 1986

ASTM Standard Technical Publication 925
Cary L. Perket, Environmental Engineering
and Management, editor.

ASTM Publication Code Number (PCN) 04-925000-16

1916 Race Street, Philadelphia, Pa. 19103

Library of Congress Cataloging-in-Publication Data

Quality control in remedial site investigation.

(ASTM special technical publication; 925)
"ASTM publication code number (PCN) 04-925000-16."
Includes bibliographies and index.
1. Hazardous wastes—Testing—Congresses.
2. Factory and trade waste—Testing—Congresses.
3. Quality control—Congresses. I. Perket, Cary.
II. ASTM Committee D-34 on Waste Disposal. III. Series.
TD811.5.Q35   1986   363.7'28   86-25873
ISBN 9-8031-0451-0

NOTE

The Society is not responsible, as a body,
for the statements and opinions
advanced in this publication.

Printed in Ann Arbor, MI
November 1986

Foreword

This publication (STP 925) contains papers presented at the symposium on Quality Control in Remedial Site Investigations, held in New Orleans, Louisiana on 8–9 May 1986. The symposium was sponsored by ASTM Committee D-34 on Waste Disposal. Cary L. Perket, Environmental Engineering and Management, served as symposium chairman and editor of this publication.

Related
ASTM Publications

Hazardous Solid Waste Testing: First Conference, STP 760 (1981), 04-760000-16

Hazardous and Industrial Solid Waste Testing: Second Symposium, STP 805 (1983), 04-805000-16

Hazardous and Industrial Waste Management and Testing: Third Symposium, STP 851 (1984) 04-851000-16

Aquatic Toxicology and Hazard Assessment: Eighth Symposium, STP 891 (1985), 04-891000-16

Rationale for Sampling and Interpretation of Ecological Data in the Assessment of Freshwater Ecosystems, STP 894 (1986), 04-896000-16

A Note of Appreciation to Reviewers

The quality of the papers that appear in this publication reflects not only the obvious efforts of the authors but also the unheralded, though essential, work of the reviewers. On behalf of ASTM we acknowledge with appreciation their dedication to high professional standards and their sacrifice of time and effort.

ASTM Committee on Publications

Contents

Introduction

The confidence with which one can propose a technical solution to any problem is directly related to the reliability of the information that is available to define the problem. For that reason, the topic of <u>Quality Control in Remedial Site Investigations: Hazardous and Industrial Solid Waste Testing</u> is of great importance to the current major effort being undertaken to find solutions to correct existing or potential threats to human health or the environment or both.

This symposium provides a state-of-the-art review of current methods to collect, prepare, and analyze samples. It also provides, for the first time, the results of a massive effort to review the quality control/quality assurance of the laboratories performing analysis under the U. S. Environmental Protection Agency's (USEAP) contract laboratory program. This symposium contains papers reporting on this program, which is believed to be the largest undertaking ever to evaluate the performance of multiple laboratories using the same analytical techniques. The multi-laboratory data provide an unequaled opportunity to gain insight into the reliability of analytical methods.

The editor would like to express his appreciation to the U. S. Environmental Protection Agency for its contribution to the program. The assistance of Mr. Gareth Pearson (USEPA Las Vegas, NV) in identifing appropriate USEPA staff and contractors for various topics was especially helpful. Finally, I would like to thank the staff of ASTM for their diligent effort in expediting this special technical publication.

Cary L. Perket
Environmental Information Ltd.
Minneapolis, MN 55435
Symposium chairman and editor

Mark S. Henne and Walter W. Meinert, P.E.

QUALITY CONTROL CONSIDERATIONS IN THE SELECTION OF WELL DESIGN, DRILLING PROCEDURES, AND WELL CONSTRUCTION MATERIALS

REFERENCE: Henne, M.S., and Meinert, W.W., "Quality Control Considerations in the Selection of Well Design, Drilling Procedures, and Well Construction Materials," Quality Control in Remedial Site Investigation: Hazardous and Industrial Waste Testing, Fifth Volume, ASTM STP 925, C.L. Perket, Ed., American Society for Testing and Materials, Philadelphia, 1986.

ABSTRACT: Groundwater monitoring wells are used extensively in the United States to detect and monitor plumes of contamination. Only with proper installation can the environmental scientist ensure that representative water samples are collected and that well installation will not contaminate "clean" aquifers. Options for drilling methods and well materials are discussed and their strengths and weaknesses are weighed.

KEYWORDS: Wells, monitor, drilling, decontamination techniques, casing materials.

Groundwater is a valuable resource in the United States and throughout much of the world. It supplies water to nearly one-half of the nation's population and accounts for about two-thirds of the fresh water resources of the world.

In the last decade, public awareness of groundwater contamination has emerged from relative obscurity to become the major environmental topic and area of legislation in the 1980's. Groundwater monitoring wells are increasingly common in industrial areas, around landfills and at many other potential sources of groundwater contamination. Data obtained from these wells are critical in identifying contamination problems and developing remedial actions. The data are frequently inaccurate, not because of laboratory error in analysis, but rather due to poorly constructed wells which yield nonrepresentative water samples. Improper monitoring well construction can also exacerbate an existing groundwater problem by allowing contaminated water into uncontaminated areas.

Monitor wells and soil borings required to install them serve four essential purposes: (1) to detect and identify contaminants in the

Mark Henne and Walter Meinert are consulting hydrologists at FTC&H Consulting Engineers, 6090 E. Fulton, Ada, Michigan 49301.

groundwater; (2) to measure the concentrations of these contaminants; (3) to characterize the geologic environment through which these contaminants migrate; and (4) to determine the direction of groundwater flow. Questions that must be asked to identify the most suitable well installation procedure are: (1) what is the expected geology of the site and what drilling options are there fore the particular geographic environment?; (2) what are the constituents to be monitored and how might these constituents react with drilling muds and well materials?; (3) at what concentrations do the parameters to be monitored exist in the groundwater, what are the desired detection limits for monitoring and how might well materials and installation procedures affect the reliability of the water quality data.

Answers to these questions must be considered in designing and constructing the most appropriate monitoring system. The purpose of this paper is to help answer these questions, to help understand the options available to the environmental scientist in selecting suitable types of well materials, and to help choose the best method of well installation.

WELL DRILLING METHODS

Several methods of installing groundwater monitoring wells are available to the environmental scientist. In choosing the best method, the site geology, along with size and type of well materials, are critical.

In most parts of the world, the earth has a layer of overburden, varying in thickness from several inches to several thousand feet, composed of unconsolidated sand, gravel, clay and silt, with interbedded rocks and boulders. This layer is deposited over bedrock of harder consolidated materials such as sandstone, limestone, granite and basalt. Groundwater is found in both the unconsolidated overburden and the underlying consolidated bedrock, each of which may be monitored. Each formation presents unique problems when installing monitoring wells. Methods that work successfully in one material are frequently ineffective or too costly to use with another.

A required well material may influence available drilling methods. Well casings and screens constructed from PVC or thin-wall stainless steel are best installed using rotary or auger drilling methods; conversely, cable tool drilling requires the use of black steel or galvanized steel casing. Often, a combination of two or more drilling techniques is used to complete monitoring wells.

Commonly used drilling methods are outlined below. Split spoon sample collection, often used to accurately identify changes in lithology, is also discussed.

Auger Drilling

Because it is fast, auger drilling is one of the most popular and common methods used to install monitoring wells throughout much of the United States. Borings are drilled using 10 to 30-centimeter diameter

screw augers. As the augers rotate and advance in the borehole, cuttings are scraped from the bottom and sides of the borehole and
brought to the surface (Figure 1). The technique is suitable only
for installing wells in unconsolidated materials. With appropriate
equipment in a favorable geologic setting, wells as deep as 50 meters
can be installed.

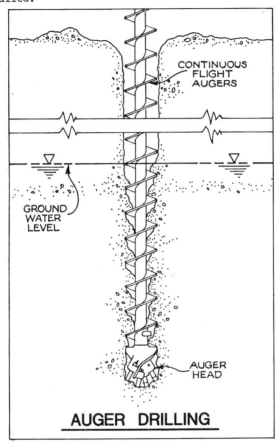

FIGURE 1: SOLID STEM AUGER DRILLING. NOTE THE COLLAPSE OF THE FORMA-
TION WHEN DRILLING IN LOOSE SAND AND GRAVEL BELOW THE WATER TABLE.

A general description of the lithology can be obtained from the
cuttings; however, more detailed descriptions can be determined by
collecting soil with split spoon samplers. Part of the difficulty in
accurately logging an auger borehole is that drill cuttings are sometimes mixed with several layers as they come to the surface. This
also creates a possibility of cross-contamination, although serious
problems are rare.

Two types of augers are used for drilling: solid stem augers
which have auger flights wrapped around a solid core; and hollow stem
augers with the flights wrapped around a heavy-duty hollow pipe.

Using hollow rather than solid stem augers can make well construction easier. Drilling can be interrupted at a selected depth, soils can be sampled through the bottom of the augers, and temporary wells can be installed for water sampling, all without removing the augers from the hole (Figure 2). This saves much time and maximizes the information that can be obtained from each borehole.

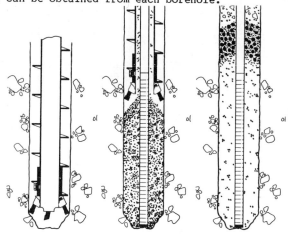

MONITORING WELL INSTALLATION
THROUGH HOLLOW STEM AUGER

FIGURE 2:   HOLLOW STEM AUGER DRILLING WITH MONITOR WELL INSTALLATION
A.   DRILLING.
B.   PVC WELL INSTALLED THROUGH CENTER OF AUGERS, AUGERS PARTIALLY WITHDRAWN AND THE GRAVEL PACK INSTALLED.
C.   AUGERS REMOVED WITH WELL IN PLACE INCLUDING BENTONITE PELLET SEAL.

Because auger drilling is so popular, new equipment is continually being developed to ease the collection of water and soil samples while drilling. Core samples several feet long can be collected in clear plastic tubes for soil indentification, and groundwater can be sampled directly through a screened auger from intermediate permeable zones without actually installing a well.

Rotary Techniques

Rotary drilling techniques are popular when drilling borings deeper than 30 meters or when drilling into consolidated (rock) formations. Rotary drilling, greatly improved by the petroleum industry, has been adapted in recent years to water well drilling.

Drill bits made of several conical roller gears are attached to the bottom of hollow drill pipe, which is rotated at approximately 30 to 70 revolutions per minute in the drill hole (Figure 3). Drilling mud formed from either clear water or water mixed with bentonite and other additives is continuously pumped through the center of the hollow drill pipe to the drill bit. The mud lubricates the bit and carries drill cuttings to the surface along the outside of the drill pipe. The fluid is then discharged to a pit where gravity separates the drill cuttings and from which mud is recirculated. With a suitable drilling rig, rotary drilling allows virtually unlimited depth capacity.

Geologic formations can be identified by inspecting drill cuttings as they emerge from the hole, although fine silt and sand are difficult to distinguish in the drilling mud. More accurate samples can be obtained by removing the drilling rod and bit and sampling with a split spoon sampler at the base of the borehole.

Cross-contamination of aquifers is possible because of the constant recirculation of the drilling mud through the borehole. This can, however, be minimized by using a thick bentonite slurry which coats the borehole and prevents migration of contaminated materials into the drilling mud, or from the drilling mud into the aquifer.

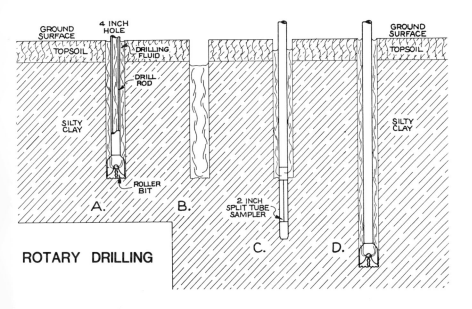

FIGURE 3:   ROTARY DRILLING WITH SPLIT SPOON SAMPLE COLLECTION

A. DRILLING OF BOREHOLE
B. BOREHOLE AFTER BIT AND RODS HAVE BEEN REMOVED
C. SPLIT SPOON SAMPLE COLLECTION
D. DEEPENING OF BOREHOLE

The possibility of contamination can also be reduced by installing casing through upper contaminated zones before continued drilling into deeper uncontaminated zones. Bentonite fluid may absorb metals and interfere with monitoring for other parameters. It can be difficult to remove from the formation during well development, but the use of trisodium phosphate helps alleviate this difficulty. Although rotary drilling is very fast, equipment costs and setup time make it expensive when installing shallow monitoring wells.

Cable Tool Drilling

Cable tool drilling, also called percussion or spudder drilling, is a very old technique, having first been used by the Chinese around 600 B.C. to install water wells. The hole is drilled by dropping a heavy, chisel-like drill bit to the bottom of the borehole to loosen unconsolidated materials and break up consolidated rock (Figure 4). Drill cuttings are removed from the hole using a bailer equipped with a check valve at the bottom. In unconsolidated formations, the hole is kept open by driving casing as soils and rock are removed.

The cable tool method has several advantages. It can be used in both consolidated and unconsolidated formations and allows continuous review of formation materials by inspection of the drill cuttings. Because the borehole is continuously cased, the possibility of mixing contaminated and uncontaminated water is minimized. The casing forms a very tight seal the full length of the well, thus making grouting unnecessary. Its major disadvantage is that it is very slow, often requiring several days to install even shallow borings.

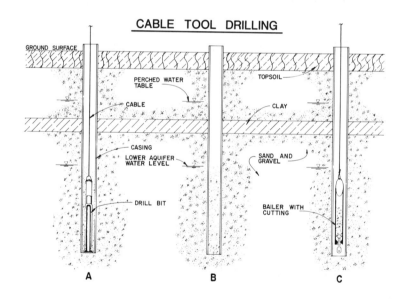

FIGURE 4:   CABLE TOOL DRILLING

A.   DRILLING OF THE HOLE AND DRIVING OF THE CASING.
B.   CASING FILLED WITH WATER CUTTINGS PARTIALLY IN SUSPENSION.
C.   REMOVAL OF CUTTINGS WITH A BAILER.

Downhole Air-Hammer Technique

With the downhole air hammer technique, casing is driven with the hammer as cuttings are blown from the well by the discharged air. The method works very well in sands or brittle consolidated materials.

The air-hammer technique is useful in quickly installing borings without the use of water or drilling mud. Formation materials can be identified as they are discharged from the hole and, if desired, split spoon samples can be collected through the casing. Steel casing materials must be used; however, permanent wells can be constructed of other materials and the casing removed.

The Advance Casing Technique

The recently developed advance-casing technique combines the advantages of continuously casing a borehole with much of the speed of rotary drilling. A specially constructed rotary bit is keyed into the bottom of drill casing. When the drilling begins, the bit expands outward to cut a hole slightly larger than the diameter of the casing. Water or drilling fluid is circulated through the center of casing to the bit and flows along the outside of the drill casing to the surface. Once the borehole has been drilled to the desired depth, a permanent well is installed through the casing which is then pulled from the borehole for reuse.

The chief advantage of the advance casing technique is that it is quicker than cable tool drilling, yet unlike rotary drilling, well screens can be temporarily set for collection of intermediate water samples. A general identification of formation materials can be made by inspecting the drill cuttings; however, more definitive identification of formation materials requires collection of split spoon samples. The advance casing technique can be the most expensive method of installing monitoring wells because of the equipment requirements and difficulties that can occur while drilling. Few drillers have the equipment to use the advance casing technique.

Jetting

Installation of wells by jetting is most common in areas of the country with loose, unconsolidated geologic materials such as sand and fine gravel. Well casing and a screen equipped with a jetting point are pushed into the sands while continuously forcing formation materials out of the borehole with a stream of water (Figure 5). Drilling penetration through loose sand and gravel can be quick, however, the rate of advance slows when denser clay is encountered. Drilling can be halted completely by a large stone or rock.

One advantage of jetting is that it is inexpensive and can be quick. The method can, however, allow cross-contamination of aquifers in a multi-aquifer environment and usually precludes installation of a bentonite seal between aquifers. Reasonably good descriptions of formation materials can be obtained by inspecting the drill cuttings and noting the rate of advance which varies with the density of the formation. Because of its disadvantages, however, it is rarely used for monitoring well installation.

Hollow Rod Method

The hollow rod method is a modification of the cable tool method. A chisel bit equipped with a check valve is attached to the end of one-inch pipe and dropped at the bottom of a continuously cased borehole. The dropping action loosens formation materials and forces water and cuttings up the one-inch pipe to the surface. Casing is intermittently driven into the formation to hold the borehole open. On reaching the desired depth, a screen is installed through the casing.

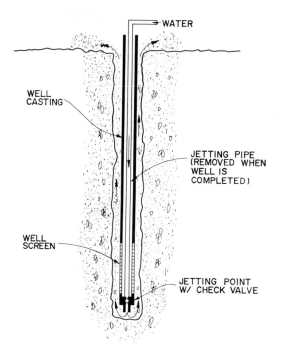

JETTING METHOD

FIGURE 5:   JETTING METHOD

WATER IS PUMPED THROUGH A PIPE TO WASHDOWN FITTING AT THE TIP OF THE SCREEN.   RETURN WATER CARRIES CUTTINGS TO THE SURFACE.   ONCE THE WELL IS IN ITS PROPER PLACE, THE JETTING PIPE IS REMOVED AND THE WELL IS DEVELOPED.

The hollow rod method has advantages over the cable tool technique in that it is often able to penetrate the formation more rapidly. The problems with the hollow rod method include difficulty in installing a bentonite seal between formations and that it is more time consuming than auger or rotary methods. As with the cable tool method, the casing forms a tight seal with the borehole, preventing vertical contaminant migration.

Split Spoon Sampling

Split spoon samples are often collected to more accurately determine geologic conditions. A 3.8-centimeter diameter sampling barrel, attached to the end of well drilling rods, is driven into the base of the borehole. Soils are pushed into the barrel and held as the sampler is pulled from the borehole. The split spoon sampler can then be separated lengthwise (from which it derives its name) and soil samples removed for classification or chemical analysis (Figures 6 and 7).

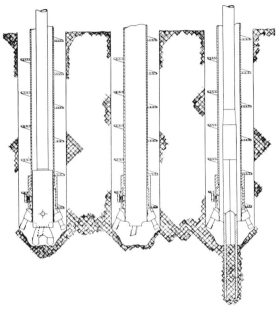

A   B   C
SPLIT-SPOON SAMPLE COLLECTION
THROUGH HOLLOW STEM AUGERS

FIGURE 6:   SPLIT SPOON SAMPLE COLLECTION THROUGH HOLLOW STEM AUGERS

A. DRILLING THE BOREHOLE WITH A TOOTHED PLUG IN THE CENTER OF THE AUGERS
B. THE PLUG IS REMOVED
C. THE SPLIT SPOON SAMPLER IS HAMMERED INTO THE SOILS AHEAD OF THE AUGERS

SELECTION OF WELL MATERIALS

Many well materials are available for installing monitoring wells. Selection of the proper materials depends on the chemical characteristics of the groundwater, the chemical parameters to be monitored, and the type of drilling technique used. Monitor wells today usually require specialized materials not commonly used in residential or industrial production wells. Figure 8 shows typical construction of a groundwater monitoring well.

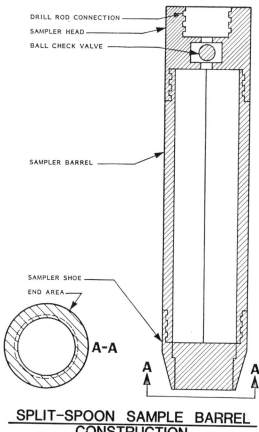

DRILL ROD CONNECTION
SAMPLER HEAD
BALL CHECK VALVE

SAMPLER BARREL

SAMPLER SHOE
END AREA

A-A

SPLIT-SPOON SAMPLE BARREL CONSTRUCTION

FIGURE 7:   CONSTRUCTION OF SPLIT SPOON SAMPLE BARRELS

BARRELS ARE GENERALLY 45 TO 61 CENTIMETERS LONG AND CONSIST OF TWO INTERLOCKING HALVES.   TO GAIN ACCESS TO THE SAMPLES, THE END CAPS WHICH HOLD THE BARREL HALVES TOGETHER ARE REMOVED AND THE BARREL IS "CRACKED" OPEN.   TUBES CAN BE INSTALLED IN THE BARREL SO THAT SAMPLES CAN BE REMOVED UNDISTURBED.

A well screen serves as a filter, keeping formation materials out while allowing water to pass. Steel screens consist of spirally wound wire surrounding and electrically welded to a number of longitudinal rods. PVC and teflon screens are formed from slotted plastic pipe. Screens of various slot widths can be used to correspond with formation grain sizes for maximum water production and minimum sand entry. Well screens can either be attached directly to the end of the casing before the well is installed in the borehole, or slid through the center of the casing after the well is installed.

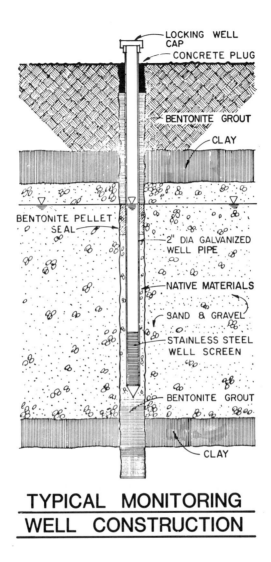

TYPICAL MONITORING WELL CONSTRUCTION

FIGURE 8:   TYPICAL MONITOR WELL CONSTRUCTION

THE WELL IS MADE OF A STAINLESS STEEL SCREEN ATTACHED TO GALVANIZED PIPE.   THE BOREHOLE THROUGH THE UPPER AND LOWER CLAYS HAS BEEN GROUTED TO PREVENT VERTICAL MIGRATION OF CONTAMINANTS AND AS AN EXTRA PRECAUTION, BENTONITE PELLETS WERE PLACED AT THE WATER SURFACE.   A CONCRETE PLUG HOLDS THE WELL IN PLACE AND STOPS SURFACE WATER FROM FLOWING THROUGH THE ANULUS BETWEEN THE CASING AND BOREHOLE.

Casing and screens are seldon necessary in consolidated rock formations where water flows to the well through interstitial pores, cracks, fissures or solution channels. This allows maximum ground-water flow into the well, minimizes the potential for altering the groundwater chemistry due to reactions of the water well screen, and reduces costs of material. Water well screens need only be installed in consolidated material if formation collapse is a possibility, or if monitoring is to be restricted to a particular depth.

Polyvinyl Chloride (PVC)

Because of its low cost and easy installation, PVC casing and well screens have been extensively used to construct monitoring wells. PVC is inert to chemical reaction in nearly every natural environment. Demands of the monitoring well industry have resulted in development of flush threaded casing and well screens to augment the more common solvent joints routinely used in the plumbing industry. Flush threaded PVC joints allow monitoring wells to be installed through the center of hollow stem augers and eliminate the possibility of solvents leaching from the joint adhesives into the groundwater being sampled.

Use of PVC materials can be limited. PVC has the potential to absorb and desorb organic chemicals, thus, if trace organic chemical concentrations are to be monitored, other well materials are better.

PVC is the weakest casing material and cannot be used when excessive stress is placed on it during well installation. PVC is not used directly with cable tool, jetting or hollow rod methods although PVC monitoring wells can be placed inside a cable tool or hollow rod well, after which the steel drive casing is removed and salvaged.

Galvanized Casing

Monitor wells are frequently constructed from five-centimeter diameter galvanized casing equipped with stainless steel well screens. Galvanized casing can be superior to PVC because it is inert to organic chemicals and more durable if the well must be driven into the formation. The galvanic coating inhibits rust formation which otherwise serves as a receptor for organic chemicals and decreases the life of the well. Although no known interferences in heavy metal analysis have been reported due to the use of galvanized water well casing, PVC casing is preferred for heavy metal monitoring.

Black Steel Pipe

Until recently, almost all water wells were constructed using black steel pipe. Today, black steel pipe is generally used when wells are at least four-inches in diameter and the boring is drilled with cable tool techniques. Because of its composition, black steel is suited to the excessive pounding and driving of the cable tool method. It is less preferred for monitoring well installation because it rusts after installation; however, it is often used when wells must be installed deeper than 30 meters. It is most suitable when rust will not interfere with chemical analyses and corrosion is not a significant factor such as when monitoring high level organic and inorganic compounds.

Stainless Steel Casing

Low chemical detection limits now attainable using GC/MS methods occasionally require that stainless steel well casing and screens be installed. Stainless steel is inert to virtually all contaminants. Its primary disadvantage is its high cost.

To help reduce cost, stainless steel casing is generally thin walled and thus lacks the ability to withstand pounding and driving required for cable tool, jetting or hollow rod methods. It is more commonly installed through the center of hollow stem augers, or after drilling the borehole with rotary methods.

Teflon

Teflon is considered the most inert well material, but because it is expensive, it is used only where no chemical interferences can be tolerated. It is structurally similar to PVC, so it is usually installed through hollow stem augers or after drilling a borehole with rotary techniques. Teflon is very soft and can be easily scratched or marred providing places for residual contaminants to adhere. Unless prices decrease, teflon will probably not be extensively used in water well construction.

MONITOR WELL GROUTING

When drilling a hole using rotary, auger or jetting methods, the final borehole diameter is larger than the well casing. If allowed to stand open or filled with permeable sand or gravel, the borehole can serve as a conduit between aquifers, and allow contaminated water to enter uncontaminated aquifers. To prevent this, the annulus between the water well casing and the borehole wall is generally grouted with bentonite, cement or a bentonite/cement mixture. An additional seal can also be obtained by using bentonite pellets which quickly sink to the base of the borehole or the top of the gravel pack and swell in contact with water. These techniques, which are effective in sealing the borehole, are described below.

Bentonite Grout

Bentonite has not only been used as the major component in drilling mud, but a bentonite slurry mixture also acts as an effective borehole seal after the well is completed. Bentonite is a clay mineral that swells and produces a slurry mixture when mixed with water. Its molecular structure is aluminosilicate sheets bonded through cation bridging. While drilling, the high-density mud carries cuttings to the surface where they are discharged to a mud pit. Constant movement keeps the bentonite clay molecules or sheets in suspension, but when agitation stops, the clay molecules begin to layer as cards would when dropped on a table. This layering creates a barrier to water moving vertically through the borehole.

Dry bentonite can be purchased as fine powder or pellets. Water molecules enter between the aluminosilicate sheets causing the bentonite to swell to as much as ten times its original size. If confined to a small area of the borehole, the swelling creates a highly impervious clay plug that halts vertical groundwater movement.

Although the clay plug created from bentonite pellets becomes a very solid mass, a seal created from bentonite slurry retains much of its semiliquid characteristics. If used alone, the slurry can be easily agitated with slight movements of the monitoring well casing. A concrete plug should be installed in the upper two to five feet of the borehole to hold the casing in place and prevent agitation.

Cement Grout

Cement grout is often used to seal the annulus, particularly after setting casing in a hole drilled with rotary methods. To install cement grout, a tremmie pipe is installed next to the well casing to the bottom of the area to be grouted. Cement grout is then injected through the pipe, displacing the drilling mud as it fills the annulus between the casing and borehole. A collar placed around the well casing, called a packer, is frequently used to limit the bottom of the grouting area. After hardening, the cement retards movement of groundwater between the borehole and the casing. Cement is more permeable to groundwater than bentonite, thus it is sometimes unsuitable; however, it is rigid and allows for better integrity around the well casing. After the water well casing has been cemented in place in a rotary-drilled boring, the well can be deepened by drilling through the base of the borehole.

Bentonite Cement Mixtures

Bentonite and cement are often mixed to create a slurry for grouting. After setting, the grout is only slightly weaker than pure cement, and little more permeable than bentonite. Variations in the mix can enhance the structural strength or the impermeability of the grout.

MONITOR WELL DEVELOPMENT

After a soil boring has been drilled and a monitoring well installed, significant concentrations of silt and drilling mud remain in the well and formations surrounding the well screen. These residual materials are removed by developing the well with compressed air, water or mechanical agitation, sometimes assisted with chemicals.

Well development is important for two reasons: it maximizes the production capacity of the well and removes foreign particles which might otherwise contaminate water samples.

The amount of well development required for a given well depends greatly on the type of formation screened.   When a well has been drilled using auger drilling techniques and is screened in a gravel or coarse sand, well development may take only a few minutes. Conversely, wells screened in silty sands can take hours or days to completely remove residual silt from the water.   The use of chemical additives such as trisodium phosphate frequently facilitates water well development.

Air Development

Air development is the most common method of monitoring well development.   A large-capacity air compressor pumps air to the bottom of the well to force water out of the screen and casing.   If the air volume is properly controlled, water flowing into the well constantly replaces the water pushed out by the compressed air.   The air column also lifts water up the casing and allows it to drop again.   This surging action breaks fine soil particles free of coarser particles, washing them away.   When developing a well with air, it is important to install a filter to remove oil from the compressed air.   Compressors frequently volatilize oil and spray it into the discharged air.

Mechanical Surging

Mechanical surging is often the most efficient means of development, particularly when large diameter monitoring wells have been installed.   A surge block is lowered to the screened interval of the well and used to churn water in the well.   This surging of water through the screen and formation removes finer particles from the formation.   Silt-laden water is then pumped or bailed from the well.

Chemical Additives

Trisodium phosphate (TSP) and other inorganic surfactants are sometimes used in conjunction with other methods to clean residual drilling materials from the well casing and dissolve and remove bentonite from the formation.   TSP chemically destroys the sheet structure of the bentonite mud and other clays, making them more soluble in water.   This greatly reduces the time required to develop the wells.   Residual manufacturing oils not completely washed from the casing and screen before installation will also be removed.   Care must be taken to ensure that all residual TSP is evacuated after development to stop it from dissolving the bentonite grout seal and prevent sample contamination.

DECONTAMINATION PROCEDURES

Analytic techniques have improved dramatically in the last decade and now many organic contaminants can be detected at concentrations as low as one ng/l; heavy metals are frequently detectable at concentrations of one ug/l or less. At these concentrations, the danger of introducing contaminants to an exploration site is always present and must be minimized to prevent cross-contamination with residual soilds from previous investigation sites where the same drilling equipment was used, or by introducing contaminants from well construction materials. Prevention of contamination of monitoring wells must be considered at all phases of water well construction, from the initial soil boring stages to the final water sampling and water level measuring stages.

Steam Cleaning

The simplest and most effective way of preventing cross-contamination is to clean equipment between borehole sites, either with steam (the most efficient) or high pressure washing. The pressurized steam frees residual soil materials, washes them from the augers, and strips organics from the metal surface. Surfactants are added to the makeup water to more completely remove oil and grease from the drilling equipment.

Galvanized or black steel casing or well screens should always be steam cleaned, as these materials have residual manufacturing oils on their surfaces when they are shipped from the factory. Steam cleaning, aided by surfactants or organic solvents such as acetone or isopropanol, effectively strips manufacturing oils from the surface. Although plastic pipe usually contains no residual oils, steam cleaning is still advisable to remove residual contaminants that might be picked up when shipped. Some care must be taken, however, when cleaning these materials since extreme heat can deform plastics.

Organic Chemical Rinses

When water samples are to be analyzed for trace organic chemicals, it may be necessary to more thoroughly clean equipment and well materials. In such cases, pesticide-free organic solvent such as hexane or isopropanol may be more useful to strip additional residual contaminants. It is important, however, that sufficient solvents be used to completely wash contaminants away from the equipment. Care must be taken when using hexane and other solvents to prevent fires and/or injury to those using the chemicals. As a further precaution, the equipment may be wrapped in aluminum foil during transport from the decontamination area to the active work site.

Heat Treatment

Heat is only occasionally used to remove residual organic contaminants. Equipment such as augers, bits and wrenches are stripped of organics using an open flame. This is less preferred for safety reasons.

Sand Blasting

Sand blasting is sometimes used to strip soil materials from augers, bits and tools. After sand blasting, equipment is normally decontaminated using a steam cleaner and solvent rinses.

CONCLUSION

Water scientists are now being required to install wells in a multitude of environments, taking into consideration numerous chemical and physical characteristics of the soils and groundwater contamination. All of these factors influence the selection of drilling techniques, well installation methods, well materials, and the requirements for grouting and decontamination of materials and equipment. Because of the many scenarios that the water scientist encounters, it is impossible to prescribe a single method for all sites. This paper serves as a guide to the options that are available.

ACKNOWLEDGEMENTS

The authors acknowledge Central Mine Equipment Company for providing Figures 1, 2, 3, 6 and 7; Sharon Ellis, Esq. for her editorial assistance; and the staff of FTC&H, particularly Stephanie Eliason and Les Hewitt, for technical help in preparing the manuscript.

BIBLIOGRAPHY

Bauwer, Herman, Groundwater Hydrology, McGraw-Hill Book Company, New York, pp 160-180.

Burklund, P.W. and Raber E., "Method to Avoid Ground Water Between Two Aquifers During Drilling and Well Completion Procedures", Ground Water Monitoring Review, Vol. 3, No. 4, Fall 1983, pp 48-55.

Curran, C.M., and Tomson, Mason B., "Leaching of Trace Organics into Water from Five Common Plastics", Ground Water Monitoring Review, Vol. 3, No. 3, Summer 1983, pp 68-71.

Diefendorff, A.F. and Ausburn, R., "Groundwater Monitoring Wells", Public Works, Vol. 108, No. 7, July 1977, pp 48-51.

Erickson, W.A., Brinkman, J.E. and Darr, P.S., "Types and Usages of Drilling Fluids Utilized to Install Monitoring Wells Associated with Metals and Radionuclide Groundwater Studies", Ground Water Monitoring Review, Vol. 5, No. 1, Winter 1985, pp 30-33.

Fishbaugh, Timothy, "Monitoring in the Vadose Zone and Unsaturated Zones Utilizing Flouroplastic", Ground Water Monitoring Review, Vol. 4, No. 4, Fall 1984, pp 183-187.

Freeze, R. Allan and Cherry, John A., Groundwater, Prentice-Hall, Inc., Englewood Cliffs, NJ, 1979, pp 5-9.

Gibb, J.P., and Barcelona, M.J., "Sampling for Organic Contaminants in Groundwater", American Water Works Association Journal, Vol. 75, No. 5, May 1984, pp 48-51.

Gordon, Raymond W., Water Well Drilling With Cable Tools, Bucyrus-Erie Company, South Milwaukee, Wisconsin, 1958.

Johnson, T.L., "A Comparison of Well Nests Vs. Single-Well Completions", Ground Water Monitoring Review, Vol. 3, No. 1, Winter 1983, pp 76-78.

Kelly, Kevin E., "Bailing and Construction Considerations for Deep Monitoring Wells on Western Oil Shale Leases", Ground Water, Vol. 20, No. 2, March-April 1982, pp 179-185.

Perazzo, J.A., Dorrler, R.C., and Mack, J.P., "Long Term Confidence in Ground Water Monitoring Systems", Ground Water Monitoring Review, Vol. 4, No. 4, Fall 1984, pp 119-123.

Perry, Charles A. and Hart, Robert J., "Installation of Observation Wells on Hazardous Waste Sites in Kansas Using a Hollow-Stem Auger", Ground Water Monitoring Review, Vol. 5, No. 4, Fall 1985, pp 70-73.

Riggs, Charles O., "Soil Sampling in the Vadose Zone," NWWA/U.S. EPA Conference on Proceedings of the Characterization and Monitoring of the Vadose (Unsaturated) Zone, Las Vegas, Nevada, December 8-10, 1983, pp 611-622.

Rinaldo-Lee, Marjory B., "Small-Vs. Large-Diameter Monitoring Wells", Ground Water Monitoring Review, Vol. 3, No. 1, Winter 1982, pp 72-75.

Salomone, L.A., "Monitoring Well Management" Chemical Substances Control, No. 143, March 13, 1986, pp 4.

Voytek, John E. Jr., "Considerations in the Design and Installation of Monitoring Wells", Ground Water Monitoring Review, Vol. 3, No. 1, Winter 1983, pp 70-71.

Williams, D.E., "Modern Techniques in Well Design", American Water Works Association Journal, Vol. 77, No. 9, September 1985, pp 68-74.

Roy B. Evans

GROUND-WATER MONITORING DATA QUALITY OBJECTIVES FOR
REMEDIAL SITE INVESTIGATIONS

REFERENCE: Evans, R B., "Ground-Water Monitoring Data Quality
Objectives for Remedial Site Investigations," Quality Control
in Remedial Site Investigation: Hazardous and Industrial
Solid Waste Testing, Fifth Volume, ASTM STP 925, C. L. Perket, Ed., American Society for Testing and Materials, 1986.

ABSTRACT: There is a need for a system of ground-water monitoring performance criteria (data quality objectives or
DQO's), reference methods, equivalency testing protocols,
and quality assurance procedures for ground-water monitoring
in remedial site investigations. This framework should be
related to relevant ambient concentration standards through
data quality objectives which allow monitoring data to be
compared to the ambient standards.

KEYWORDS: data quality objectives, performance criteria,
ground-water monitoring, ambient concentration standards

Introduction

Before a monitoring system can be established in any medium, the
purpose of the system needs to be established. This statement of purpose has been called performance criteria in some fields of environmental monitoring, and the present EPA designation of choice is data
quality objectives. This paper is an attempt to describe what these
data quality objectives might be for remedial site investigations and to
indicate their implications for ground-water monitoring systems. The
statement of these performance criteria or data quality objectives is a
necessary first step in establishing standard monitoring methods.

Background

Although dozens of publications discuss ground-water monitoring in
varying degrees of detail, four particularly important documents are

Dr. Evans is an environmental scientist at the Environmental Research Center, University of Nevada, Las Vegas, 4505 Maryland Parkway,
Las Vegas, NV 89154.

expressly devoted to specific aspects of the subject:

"TEGD":   RCRA Ground-Water Monitoring Technical Enforcement Guidance
           Document (Draft), EPA, OSW, August 1985.

"PWG":   RCRA Permit Writer's Manual, Ground-Water Protection, 40 CFR
           Part 264, Subpart F (Draft), Geo-Trans, Inc., October 4, 1983
           (developed for OSW on contract)

"Interim   Ground-Water Monitoring Guidance for Owners and Operators of
Guide":   Interim Status Facilities: Instructions for Complying with 40
           CFR Part 265, Subpart F; EPA, OSW, 1982.

"Scalf   Manual of Ground-Water Sampling Procedures, M.R. Scalf, J.F.
Manual":   McNabb, W.J. Dunlap, R.L. Cosby, and J. Fryeberger; EPA,
           RSKERL, 1981.

The TEGD discusses the total scope of ground-water monitoring for
purposes of RCRA enforcement. It discusses the steps involved in char-
acterizing site hydrogeology, placement of monitoring wells, monitoring
well design and construction, sampling and analysis, statistical analysis of detection monitoring data, and assessment monitoring.

The PWG is another comprehensive discussion of ground-water monitoring systems for land disposal facilities permitted under 40 CFR Part
264. It contains much useful material. Major discussion topics include
evaluation of monitoring network design, evaluation of monitoring well
design and construction, evaluation of sampling and analysis procedures,
statistical analysis of monitoring data, establishing the ground-water
protection standard, detection monitoring programs, compliance monitoring programs, and corrective action programs.

The Interim Guide goes through a similar, though more cursory
discussion for interim status sites regulated under 40 CFR Part 265.
The document also discusses the use of tracers and indirect techniques
for assessing ground-water contamination, including aerial photography
and both surface-based and downhole electrical geophysical techniques.
It concludes that such techniques can reduce the costs of assessment
monitoring while adding significantly to the understanding of a contamination incident. It stresses, however, that these methods cannot completely replace the collection and analysis of ground-water samples.

The Scalf Manual is an early EPA manual specifically dealing
ground-water monitoring. Although prepared prior to the promulgation of
40 CFR Part 264, it contains detailed descriptions of specific sampling
techniques that are still useful.

Two more recent EPA publications also add materially to available
information:

"Materials   A Guide to the Selection of Materials for Monitoring Well
Guide":   Construction, M.J. Barcelona, J.P. Gibb, and R.A. Miller,
             Illinois State Water Survey, EPA-600/2-84-024, May 1983.

"Practical   Practical Guide for Ground-Water Sampling," M.J. Barcelo-
Guide":   na, J.P. Gibb, J.A. Helfrich, and E.E. Garske, Illinois
             State Water Survey, EPA/600/2-85/104, September 1985.

The Materials Guide discusses the chemical compatibility of well construction materials with various kinds of contaminated ground waters and
describes tests to determine for specific contaminants their leaching
from or removal by ground-water monitoring systems components. The
Practical Guide is a logical discussion of the development of an overall
ground-water monitoring program and network, covering largely the same
topics as the TEGD, PWG, and Interim Guide, but emphasizing ground-water
chemistry and including some recommendations concerning construction
materials.

Some limited conclusions can be drawn about gaps in existing guidance. While there exists an impressive body of literature on ground-
water monitoring, there appear to be no established monitoring system
performance specifications or data quality objectives to quantitatively
relate errors in measured concentrations to ambient concentration standards for specific hazardous constituents. Neither are there any protocols for determining the effects either of specific components or of the
total monitoring system on measured concentrations. Existing EPA
ground-water monitoring guidance is concerned primarily with the mechan-
ics of getting samples out of the ground and into the laboratory. Only
recently has there been research directly concerned with measuring the
impact of well location, drilling, construction purging, and sampling on
the chemistry of the sample as it arrives in the laboratory. There has
been very little effort devoted to systematic investigations of the
effects of leaching and/or adsorption by components of the ground-water
monitoring system on collected samples.

Standard Monitoring Methods

The development of standard environmental monitoring methods has
usually followed a path similar to that illustrated in Figure 1, leading
from an ambient concentration standard to a reference method for the
pollutant of interest and subsequently to equivalent methods. The
normal objective is the enforcement of an **ambient concentration standard.** This is generally a health-based maximum concentration of a
particular toxic chemical. The ambient standards to be enforced lead to
performance criteria for the total ambient monitoring system. Simply
put, the ambient monitoring system must measure ambient concentrations
of the toxic substance "well enough" to allow reliable comparisons
between the measurements and the pertinent health-related ambient standard. The performance criteria define "well enough" by specifying
accuracy, precision, bias, sensitivity, minimum detectable concentrations, representativeness, completeness, and comparability. For purposes of the present discussion, these performance criteria can also be
considered "Data Quality Objectives." Based on the monitoring system
performance criteria, a set of **reference methods** for monitoring ambient
concentrations is selected to meet the performance criteria. Following
the establishment of reference monitoring methods, **equivalency testing
procedures** are usually defined to allow the testing of new, candidate
monitoring methods. Data from equivalency testing are reviewed by the
EPA, and, if satisfactory, lead to recognition of the candidate method
as an **equivalent method.**

To support the use of reference and equivalent monitoring methods
in field monitoring programs, a framework of quality assurance procedures is usually essential. These include audit procedures for complete

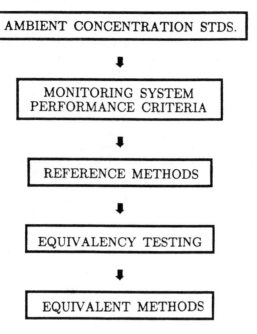

Fig. 1 -- Development of standard
ambient monitoring methods

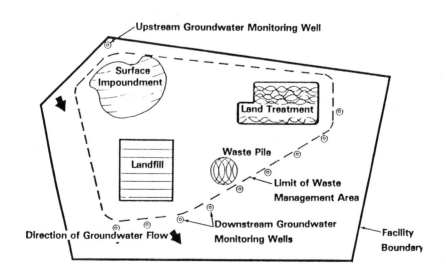

Fig. 2 -- Facility overview

monitoring systems, as well as individual components.

In the following discussion, a framework for development of standard ground-water monitoring methods and quality assurance procedures, analogous to that which exists for ambient air quality monitoring, is described. No set of environmental concentration standards applicable to CERCLA have been promulgated. However, it is reasonable to expect that eventually, a set of ambient concentration standards similar to that being developed for RCRA will be established for CERCLA remedial action. A logical possible approach to setting these standards would be to use available guidance for RCRA, since CERCLA and RCRA would presumably not be inconsistent. As a starting point, the Ground-Water Protection Regulations of RCRA are reviewed, and the maximum concentration limits (MCL's) and ACL's are listed. The framework for an equivalency program is outlined.

RCRA Ground-Water Protection Regulations

The ground-water protection requirements of 40 CFR Part 264, Subpart F establish a three-stage program to detect, evaluate, and correct ground-water contamination. The program must be complied with throughout the active life of the facility and for 30 years after closure. The first stage of the ground-water monitoring and response program is a detection monitoring program, which requires the permittee to install a ground-water monitoring system (including both upgradient and downgradient wells) at the boundary of the waste management area (see Figure 2). The permittee must monitor the ground water in the uppermost aquifer for parameters that would indicate whether a leachate plume has reached the waste boundary. If a plume is detected, a second stage - a compliance monitoring program - is established. The compliance monitoring program monitors the concentration of specific hazardous constituents that are reasonably expected to be in waste disposed at the facility and that are found in the ground water.

The results of compliance monitoring are compared against a ground-water protection standard. The standard requires that hazardous constituents not exceed the following concentration limits:

(1) The background level in the ground water; or

(2) For any of the 20 hazardous constituents covered by the National Primary Drinking Water Standards (NPDWS), the maximum concentration limits for drinking water established in those regulations (unless the background levels are higher); or

(3) Alternate concentration limits (ACL's) which the permittee successfully demonstrates to be fully protective of human health and the environment.

If the standard is violated, the third stage - corrective action - is activated. Corrective action must continue until the standard is met, even if it extends beyond the 30-year post closure period. Corrective action consists of the removal of the contamination (by pumping or other means) or in-situ treatment of the hazardous constituents to restore ground-water quality (see Figure 3).

Both the design and operating standards and the ground-water moni-

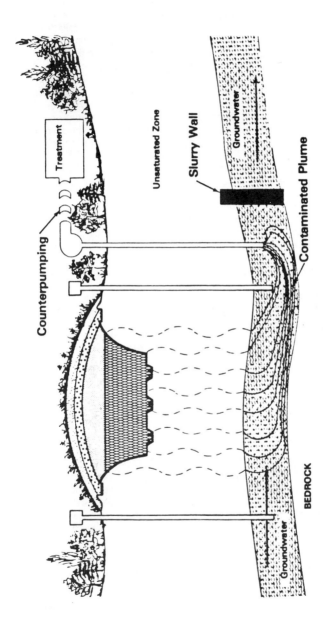

Fig. 3 -- Corrective action

toring and response program are to be implemented through the issuance of permits. In the case of the ground-water monitoring and response program, permit modifications are required when there is a need to progress from one stage of the program to the next.

Detection Monitoring

Detection monitoring, the first stage of the 40 CFR, Part 264, Subpart F ground-water protection program, is to be implemented at hazardous waste facilities where no hazardous waste constituents are known to have migrated from the facility to ground water and is designed to alert the owner/operator when a leachate plume from the facility first reaches the ground water. The program requires the permittee to install a monitoring network which includes wells located downgradient from the facility at the limit of the waste management area and wells located upgradient that are capable of providing samples representative of background water quality in the vicinity of the facility (see Figure 2). A set of **indicator parameters** is identified which is capable of detecting the arrival of the leachate plume from the facility. The concentrations of these indicator parameters are to be routinely monitored in the downgradient wells. Indicator parameters suggested for possible use under Part 264 include pH, specific conductance, total organic carbon, and total organic halogen. Should the arrival of leachate from the facility be indicated by the measurement of increased (or decreased) concentrations of any of the indicator parameters relative to natural background levels, the permittee must immediately institute a sampling program to determine the chemical composition of the leachate plume reaching the downgradient wells.

Compliance Monitoring

Whenever detection monitoring has indicated that hazardous constituents from the facility have reached ground water, the owner or operator must institute a **compliance monitoring program**. The compliance monitoring program is the second stage of Subpart F's monitoring and response program, and its requirements are intended to address the specific chemical constituents in the leachate coming from the unit. In most instances, the monitoring network established for detection monitoring will be adequate. In addition to monitoring to ensure that the ground-water protection standard is met, compliance monitoring includes routine sampling to detect any additional hazardous constituents that might enter the ground water from the facility. Should any additional constituents appear, these constituents must also be listed in the ground-water protection standard and appropriate concentration limits established.

Ground-Water Protection Standard

The **ground-water protection standard** identifies the individual constituents that must be routinely monitored and specifies the concentration of each hazardous constituent that triggers the need for corrective action measures. The first element of the ground-water protection standard is a **listing of all hazardous constituents** present in the ground water that can reasonably be expected to have come from the regulated unit. Hazardous constituents have been defined as any constituent listed in Appendix VIII of 40 CFR Part 261.

The second element of the ground-water protection standard is the specification of concentration limits for each of the hazardous constituents listed in the permit. The regulations have adopted an approach of basing, when possible, concentration limits on well-established numerical concentration limits for specific constituents (for example, the National Primary Drinking Water Standards mentioned above). Before such standards are promulgated, concentration limits for hazardous constituents are to be based on existing ground-water quality background levels.

If compliance monitoring indicates that the ground-water protection standard is being violated, the permittee must activate the third phase of Subpart F's monitoring and response program, the corrective action phase. The applicant must submit a plan for corrective action which must be approved and incorporated into the facility permit. The plan must be designed to achieve compliance with the ground-water protection standard and must be initiated and completed within a reasonable time, considering the extent of contamination. Compliance must be achieved by either removing the hazardous constituents or by treating them in place.

The regulations also require corrective action programs to include monitoring programs adequate to demonstrate the effectiveness of the corrective action measures. This will likely require the expansion of existing monitoring networks used in routine detection and compliance monitoring. Corrective action programs must be continued until compliance with the ground-water protection standard is achieved. Once compliance has been achieved, the permittee may return to a compliance monitoring program. However, if corrective action measures have continued past the normal compliance period specified in the permit, the permittee must continue to monitor to show that the ground-water protection standard is not exceeded for an additional 3 years. Table 1 summarizes the ground-water protection monitoring requirements of 40 CFR Part 264.

Table 2 attempts to summarize the above discussion for Part 264 in terms of the (ambient) ground-water concentration standards against which the measured ambient concentrations are to be compared. In detection monitoring, measured concentrations or parameter values are compared with background concentrations measured during the background monitoring period. In compliance monitoring measured values are compared with the National Primary Drinking Water Standards for the 20 parameters which have standards.

Derivation of Monitoring Performance Criteria

Ground-water monitoring system performance criteria (Data Quality Objectives) should be related to the appropriate ambient concentration standards. One possible approach is to use relationships between the ambient concentration standards and the monitoring performance criteria similar to those found in ambient air quality control. Table 3 lists performance parameter specifications for equivalent automated ambient air quality monitoring methods (from 40 CFR, Part 53) along with the ambient air quality standard(s) for each pollutant. Some of these performance parameters clearly relate to automated, continuous samplers and are inappropriate for the current state of ground-water monitoring technology; zero and span drift, and lag, rise, and fall times have no meaning for collection and analysis of discrete ground-water samples.

TABLE 1 -- Summary of RCRA Groundwater Monitoring

HAZARDOUS WASTE FACILITY PERMIT

Element		Duration	Monitoring	
			Analysis	Frequency
If no hazardous contituents in groundwater...	Detection Monitoring	Until detection or approximately 30 years post closure	For a few indicator parameters	Semi-annual
If hazardous constituents are in groundwater...	Compliance monitoring	At least equal to total operation life of facility	To compare against concentration limits for hazardous constituents	Quarterly
If hazardous constituents exceed limits...	Corrective action	As long as necessary		

TABLE 2 -- Reference Concentrations RCRA
Groundwater Monitoring

	Part 264 (New Sites)
Detection Monitoring	Background
Compliance	Permitted Maximum Concentrations
Corrective Action	Permitted Maximum Concentrations

Table 3
Performance Specifications for Equivalent Air Quality Monitoring Methods

Performance Parameter	Units	Sulfur Dioxide	Photo-chemical Oxidants	Carbon Monoxide	Nitrogen Dioxide
Ambient Air Quality Std	ppm	.03 (annual) .14 (24-hr)	.12	9.0 (8-hr) 35 (1-hr)	.05
Range	ppm	0 - .5	0 - .5	0 - 50	0 - .5
Noise	ppm	.005	.005	.5	.005
Lower Detectable Limit	ppm	.01	.01	1.0	.01
Interference Equivalent	ppm				
Each Interferant	ppm	.02	.02	1.0	.02
Total Interferant	ppm	.06	.06	1.5	.04
Zero Drift, 12 & 24-hour	ppm	.02	.02	1.0	.02
Span Drift (24-hour)					
20% of upper range limit	percent	20.0	20.0	10.0	20.0
80% of upper range limit	percent	5.0	5.0	2.5	5.0
Lag Time	minutes	20	20	10	20
Rise Time	minutes	15	15	5	15
Fall Time	minutes	15	15	5	15
Precision					
20 % of upper range limit	ppm	.01	.01	.5	.02
80 % of upper range limit	ppm	.015	.01	.5	.03

However, range, lower detectable limit, interferences, and precision are useful concepts for ground-water monitoring.

Table 4 lists the values for National Primary Drinking Water Standards. In compliance monitoring at facilities which have received permits under Part 264, measured values are compared with the MCL's and ACL's listed in the permit. Tables 2 and 4 are the ambient concentration standards identified by Figure 1 as the first step in establishing performance criteria for ground-water monitoring methods.

Ambient concentration standards are based on exposure assessment and risk assessment considerations; the intent is to ensure that a person exposed to a polluted environmental medium receives no more than an "acceptable" dose of a particular toxic substance or hazardous constituent. The "acceptable" dose is based on risk assessments to estimate the health consequences of such a dose. These exposure assessment considerations imply a specification of both sampling schedules and data analysis techniques in the designation of ambient concentration standards and reference measurement methods. For example, the air quality standard for ozone requires that ambient ozone concentrations in a locality not exceed an hourly average of 0.12 ppmv (parts per million by volume) more than once per year; this implies the continuous measurement of ozone and the compilation of hourly averages for every hour of a year. Data analysis consists of preparation of a cumulative frequency distribution from these hourly averages to estimate the probability that any hourly average exceeds 0.12 ppmv. In the case of RCRA MCL's, the national primary drinking water standards (NPDWS) must be met. The NPDWS require the annual average concentration of mercury in drinking water to be less than 0.002 mg/l. At a RCRA facility where mercury is disposed, this implies that the sampling frequency must be adequate to determine the annual average mercury concentration in ground water exiting the facility boundary. By analogy with the air pollution example, data analysis would consist of calculating this annual average.

Table 5 lists a highly tentative set of possible general performance criteria for ground-water monitoring as fractions of the appropriate ambient concentration standard. These fractions are based on similar ratios for ambient air monitoring calculated from Table 3. The suggested range of measurement is at least ten times the MCL, ACL, or NPDWS; the lower detectable limit, permissible interferences, and precision would each be of the order of 10 percent of the concentration standard. Table 5 also gives an example of the application of these criteria to monitoring for mercury. Using these general criteria, the suggested range for mercury would then be at least 0 to 0.02 mg/l; the lower detectable limit, total interference equivalent, and precision would each be equal to or less than 0.0002 mg/l. It must be emphasized that these figures should be applied to the overall ground-water monitoring system; they refer to the sum of the errors introduced through well drilling, construction, completion, and pumping, and through sample collection, preservation, transport, storage, and analysis. It must also be emphasized that the limits of the analytical methods will often offer constraints. In the case of mercury, the detection limit for Method 7470 (manual cold-vapor technique) is 0.0002 mg/l [1]. At concentrations as low as the standard of 0.002 mg/l, the precision of Method 7470 is probably much poorer than 10 percent [2]. Indeed, we can expect the minimum detectable concentration of the analytical method to be used as the ambient standard for many substances.

TABLE 5 -- Possible Performance Standards for
Groundwater Monitoring

CONTAMINANT	CONCENTRATION (mg/l)
Arsenic	0.05
Barium	1
Cadmium	0.010
Chromium	0.05
Lead	0.05
Mercury	0.002
Nitrate (as N)	10
Selenium	0.01
Silver	0.05
Floride (at 20 °C)	1.8
Endrin	0.0002
Lindane	0.004
Methoxychlor	0.1
Toxaphene	0.005
2,4–D	0.1
Silvex	0.01
Trihalomethanes	0.10
Radium	5 pCi/l
Gross Alpha	15 pCi/l
Gross Beta	dose <4 mrem/yr

TABLE 4 -- National Primary Drinking Water Standards

Performance Parameter	Possible Performance Criterion	Possible Performance Criterion for Mercury (Hg) (mg/l)
Ambient Standard	– – – – –	.002
Range	10 x MCL, ACL, or NPDWS	0 – .02
Lower Detectable Limit	10 % of MCL, ACL, or NPDWS	.0002
Noise	10 % of MCL, ACL, or NPDWS	.0002
Interference Equivalent	10 % of MCL, ACL, or NPDWS	.0002
Precision	10 % of MCL, ACL, or NPDWS	.0002

The list of parameters in Table 5 are goals or objectives for the performance of the overall monitoring system. This list may need to be augmented by more detailed supplemental criteria; for example, to ensure that the suggested interference equivalent for mercury really is less than 0.0002 mg/1, specification of drilling methods and well casing and pump materials may be necessary. To ensure that the precision criterion is met, specification of well purging procedures may be required. Hence, the overall process of developing performance criteria will be complicated and must be approached systematically.

Conclusions

There is a need for a system of ground-water monitoring performance criteria (data quality objectives or DQO's), reference methods, equivalency testing protocols, and quality assurance procedures for ground-water monitoring in remedial site investigations. This framework should be related to the MCL's, ACL's, and indicator parameter detection monitoring trigger limits which are the applicable ambient concentration standards. While a substantial body of ground-water monitoring guidance exists, little of it can be quantitatively related to ambient concentration standards. A systematic development effort is needed to establish for ground-water monitoring the type of framework indicated in Figure 1.

The critical link in this framework is the development of data quality objectives or performance criteria for remedial investigation ground-water monitoring. As suggested earlier, these performance criteria should be quantitatively related to ambient concentration standards. Probably the most relevant set of standards is the set of MCL's, ACL's, and indicator parameter background concentrations being developed for RCRA ground-water protection monitoring. In any case, standards for remedial site work should not be inconsistent with RCRA standards. Since in RCRA there could conceivably be standards not only for the substances covered by the NPDWS but also for the approximately 400 substances identified in Appendix VIII of 40 CFR Part 261, this step could be a major effort and should be coordinated with RCRA standards development. It may be necessary to develop DQO's for classes of compounds rather than individual substances. Consideration must be given not only to the actual values of the standards but also to the limits of available analytical methodology. In many cases appropriate MCL's or ACL's will not exist, and interim values will have to be used, such as the minimum detectable concentrations of the corresponding analytical methods.

A necessary, related step is a comprehensive examination of existing RCRA and Superfund ground-water monitoring data to determine the spectrum of substances most commonly observed and the ranges of concentrations encountered. This review was begun several years ago by the EPA, and it should be possible to draw some preliminary conclusions [3, 4].

For practical reasons, performance specifications for ground-water monitoring methods may be needed which are not directly related to the DQO's. A possible example is well diameter, in cases where it may be desirable to use monitoring wells for geophysical logging or other purposes which are not strictly part of sample collection. The existing body of literature contains such recommendations, and a thorough, in-depth, critical review of existing literature must be conducted to identify such issues for consideration.

REFERENCES

[1]   Office of Solid Waste, U.S. Environmental Protection Agency, Test Methods for Evaluating Solid Waste, Final Report, (SW-846) July 1982.

[2]   Methods Development and Quality Assurance Research Laboratory, National Environmental Research Center, Cincinnati, Ohio, "Methods for Chemical Analysis of Water and Wastes," (EPA-625/6-74-003) 1974.

[3]   Plumb, R. H. Jr., and C. K. Fitzsimmons, "Performance Evaluation of RCRA Indicator Parameters," First Annual Canadian-American Conference on Hydrogeology, Banff, Alberta, June 22-26, 1984

[4]   Plumb, R. H. Jr., "Disposal Site Monitoring Data: Observations and Strategy Implications," Second Annual Canadian-American Conference on Hydrogeology, Banff, Alberta, June 25-29, 1985.

R. J. Bruner, III, P.E.

A REVIEW OF QUALITY CONTROL CONSIDERATIONS IN SOIL SAMPLING

REFERENCE: Bruner, R. J., III, "A Review of Quality Control Considerations in Soil Sampling," Quality Control in Remedial Site Investigation: Hazardous and Industrial Solid Waste Testing, Fifth Volume, ASTM STP 925, C. L. Perket, Ed., American Society for Testing and Materials, Philadelphia, 1986.

ABSTRACT: Precision, accuracy, and representativeness of samples are as important in soil sampling programs as these considerations are for other quality assurance programs. Proper procedures for cleaning and preparing sampling equipment and containers, selecting sampling sites, sample collecting and handling, and laboratory quality assurance/quality control should be observed to preserve the integrity of soils analyses.

KEYWORDS: metals, procedures, quality assurance, quality control, sample collection, sample handling, sampling, soils, organic chemicals, organic compounds

INTRODUCTION

Quality assurance for soil sampling involves the same basic quality assurance considerations as other sampling activities (i.e., precision, accuracy, and representativeness). This paper addresses these considerations as they relate to the collection of soil samples for chemical analyses for trace concentrations of organic compounds and metals. Factors that will be addressed include:

1. Sampling equipment and sample containers,
2. Cleaning procedures,
3. Site selection criteria,
4. Sample collection procedures,
5. Quality control samples, and
6. Sample handling procedures.

SAMPLING EQUIPMENT AND SAMPLE CONTAINERS

The primary factor to consider when selecting sampling equipment and containers is the nature of the materials that will be in contact

Mr. Bruner III is a project manager in CH2M HILL's Hazardous Waste/Industrial Processes Group, 7201 N.W. 11th Place, P.O. Box 1647, Gainesville, FL 32602.

with the sample. The abrasive nature of soils makes material
selection more critical than in some other types of sampling, since
material may be scraped from the sampling equipment and/or containers,
thus becoming incorporated in the sample. Glass and Teflon® are inert
to most chemicals and are generally free of compounds that could
contaminate the soil during sampling; however, the structural strength
of the materials is insufficient for many sampling applications.

Plastics other than Teflon® generally contain plasticizer and
other chemical additives which can contaminate samples being analyzed
for organic compounds [1-3]. Plastics may also sorb organic compounds,
thus lowering the concentration present in the sample. When sampling
for metals or other inorganic parameters, the use of plastics is
generally acceptable but plastic should be avoided when sampling for
organic compounds.

Sampling equipment constructed of metal also presents problems
which should be considered especially when collecting samples for
metals analyses. Much of the equipment used to collect soil samples is
constructed of metal because of the structural strength required for
digging. Stainless steel is fairly inert and does not corrode, thus
preventing corrosion from being scraped from the equipment into the
sample. The disadvantage of stainless steel is the metal alloys
present in the steel. Plated and galvanized steel also presents a
problem since metals present in the coating may flake off or be scraped
off and enter the sample. Carbon steel does not contain other metals
that present a contamination problem but it corrodes readily.
Corrosion presents several problems, such as flaking, which can
contaminate the sample. Corrosion also makes cleaning and decontamination more difficult and provides sites for ion exchange and/or sorption
of organic compounds. Aluminum pans, which are handy for mixing
samples, have been reported by chemists with the USEPA, Region IV
Environmental Services Division to cause analytical problems during
metals analyses [4]. Because the aluminum is soft, it is apparently
scraped off the pans and into the sample, producing extremely high
aluminum concentrations that mask other metals during analysis.

Spoons, hand augers, shovels, and other items of manual soil
sampling equipment are generally available in stainless steel. Pans
and containers for compositing and mixing soils are available in stainless steel, glass, and aluminum. Many of the smaller items of sampling
equipment are available with Teflon® coating or can be Teflon® coated;
however, the Teflon® wears off very quickly and is, therefore, relatively ineffective. Other coating materials, such as paints or
plating, should be avoided or removed from surfaces that come in
contact with samples.

Power sampling equipment is generally constructed of carbon steel
because of strength requirements. When using power sampling equipment
such as drilling rigs or power augers, no lubricants should be used on
joints and connection, and all paints, coatings, and corrosion should
be removed from surfaces that come in contact with the sample.

CLEANING PROCEDURES

Cleaning sampling equipment and sample containers is important to

avoid cross-contamination of samples and/or the introduction of
extraneous materials into the samples. The following cleaning
procedure is recommended for sampling equipment [5, 6]:

1. Wash equipment using phosphate-free, laboratory grade
    detergent and potable water.

2. Rinse thoroughly with potable water.

3. Rinse with 20 percent nitric acid followed by another potable
    water rinse (glass, Teflon®, and plastics only).

4. Rinse with distilled or deionized water.

5. Applicable only when sampling organic compounds:

    a. Rinse with pesticide grade solvent (isopropanol is
        recommended for general usage; other solvents are accep-
        table and at times preferred depending on application).

    b. When possible, oven dry at 105°C for at least one hour;
        otherwise, allow to air dry (preferably overnight).

6. Wrap in aluminum foil until ready to use.

In some cases, especially when using power drilling equipment,
sand/abrasive blasting may be necessary to remove soil buildups and
corrosion. When required, sand/abrasive blasting is normally performed
prior to the start of sampling at a site but is not required between
samples. Steam cleaning or high pressure washing is an acceptable
alternative to the soap and water wash (Step 1 of the above procedure).

Distilled/deionized water and solvents should be stored and
dispensed from glass or Teflon® rather than plastic containers to avoid
the introduction of organic compounds. Compressed air pump sprayers
should be avoided for dispensing distilled/ deionized water and
solvents because of the rubber gaskets and lubricants used in the pumps
[5].

SITE SELECTION

Sampling locations should be selected carefully to ensure
representativeness of the sample and to avoid unintentional biasing of
samples. Unintentional biasing of a sample occurs when conditions that
are not apparent to the field personnel result in a sample that is not
typical of the normal condition in the sampling area (i.e., a sample is
collected from a spill area unknown to the sampling personnel). Intentional biasing of a sample, collection within a spill or depositional
area, is an acceptable site selection procedure since the sampling
personnel are aware of the bias and the data can be interpreted
accordingly [7, 8].

Sampling locations can be selected authoritatively, systematically, or randomly depending on the purpose of collection. Authorita-
tive or judgmental sampling site selection is generally used during
reconnaissance investigations and site screening investigations. This
procedure requires knowledge and judgment by the sampler and often

involves intentional sample biasing (i.e., collection of a sample from "hot spots" and/or an environmental sink to evaluate worst case conditions). Systematic sample site selection is normally used when attempting to determine areal extent of contamination or when evaluating spatial variations. Sampling locations are defined by a grid or coordinate system and samples are collected at preselected locations in a uniform pattern [7, 8]. Random sample site selection is used when attempting to determine statistical parameters such as mean and variance. Random site selection involves the use of a random number generator to select sampling locations based on a grid system or transect [9].

SAMPLE COLLECTION

The collection of soil samples may involve surface sampling, shallow subsurface sampling, or deep subsurface sampling. Surface and shallow subsurface samples are normally collected using manual equipment. Deep subsurface sampling requires the use of power equipment such as drilling rigs. Specific sampling methodologies are beyond the scope of this paper; however, some general considerations relating to quality control must be addressed. Available sources of specific sampling methodologies include the following:

1. Draft Water Surveillance Branch Standard Operating Procedures and Quality Assurance Manual, U.S. Environmental Protection Agency, Environmental Services Division, Athens, Georgia.

2. Preparation of Soil Sampling Protocol: Techniques and Strategies, (EPA-600/4-83-020), Environmental Monitoring Systems Laboratory, Office of Research and Development, U.S. Environmental Protection Agency, Las Vegas, Nevada.

When possible, sampling should progress from least contaminated to most contaminated to reduce the potential for cross-contamination of samples. If this is not possible, extreme care must be taken to ensure that the sampling equipment is adequately cleaned between samples and that materials on the samplers hands and/or clothing do not contaminate the sample. If two sampling personnel are available and a "clean" sample must be collected after a "dirty" sample, one individual should take the "clean" sample, while the other takes the "dirty" sample, thus avoiding possible contamination of the "clean" sample by materials from the sampling personnel's hands and/or clothing.

When subsurface samples are being collected care must be taken to avoid contaminating the sample with material from overlying strata. This involves removal of loose material which can fall into the hole prior to sampling, and being careful not to dislodge material from the side of the hole during sample retrieval. Also, the same equipment used to advance the hole should not be used for sample collection since the equipment may have been contaminated by overlying strata.

Control or background samples should be collected to provide a point of reference when evaluating data. This is especially important when evaluating metals data or data associated with other naturally occurring soil constituents. Care should be exercised in selecting the control or background sampling location so that the area selected is unaffected by activities of the site being investigated (control sample), or is representative of true background conditions (natural condition unaffected by the activities of man) [5].

Another important aspect of quality assurance is sample documentation. This includes identifying (tagging or labeling) samples, maintaining a field notebook, and tracking the samples from time of collection through analytical data reporting (chain-of-custody). At a minimum, field documentation should include sampling location, date and time of collection, sampling personnel, sampling method, depth interval sampled, any unusual observations (i.e., abnormal color, odors, etc.), and any deviation from standard procedures. Field notes should be maintained in a bound notebook using waterproof ink. All samples should be identified with a station number, date and time of collection, sampling personnel, type of preservation, and analyses to be performed. The integrity of the samples should be protected from the time of collection until the analytical results are reported. A documented chain-of-custody procedure is recommended for tracking the samples from the field to the laboratory [5, 9].

QUALITY CONTROL SAMPLES

Blanks, spikes, duplicate and split samples are all valuable tools in determining the precision and accuracy of analytical data. Blank samples are divided into several types including laboratory blanks, field (or trip) blanks, equipment blanks, and container blanks. A laboratory blank is prepared in a laboratory, coded as a sample, and delivered to the analytical laboratory. The purpose of a laboratory blank is to provide an independent check for possible laboratory contamination. Handling of the laboratory blank in the field is limited to that which is required for documentation and shipping. Laboratory blanks are kept isolated from samples and sampling equipment to avoid potential contamination except as required for shipping. Spike samples are handled in the same manner as laboratory blanks. They are prepared in a laboratory, coded as a sample, and delivered to the analytical laboratory for analysis. Handling is minimized in the field. The purpose of blind spike samples is to evaluate recoveries by the laboratory. The resulting data are used to evaluate the accuracy of the analytical data.

A field or trip blank is used to detect contamination associated with sample handling, and is very important when sampling for purgeable organic compounds. Trip blanks are prepared in the laboratory but, unlike laboratory blanks, are carried to the field and kept with the samples during sampling activities.

Equipment and container blanks are used to detect contamination associated with inadequate cleaning of sampling equipment, or sample containers. Equipment and container blanks are prepared in the field using media furnished by the laboratory. The blank media is handled as a sample using sampling equipment and containers from the lot used for the sampling activities. Water is often used instead of soil for the preparation of equipment and container blanks. The sampling equipment is rinsed with water and the rinsate is placed in a sample container for analysis [5, 8].

Split or duplicate samples are used to evaluate precision. Split samples are those that have been placed in a common container, thoroughly mixed, and then placed in a sample container. Duplicate samples are collected at the same location and time but without being mixed in a common container. Split and duplicate samples can be coded as different samples and delivered to the same laboratory or can be coded as the same sample and delivered to different laboratories. Often both types of split/duplicate samples are used during an intensive sampling effort to evaluate inter- and intra-laboratory precision and accuracy. Split and duplicate sample data are also valuable for evaluating the homogeneity of the soil being sampled and the adequacy of mixing.

The quality control samples addressed in this paper relate to field quality control and are independent of internal laboratory quality control procedures. Under normal circumstances, approximately ten to fifteen percent of the samples submitted to the laboratory should be field quality control samples. These samples should be uniformly divided between equipment blanks, container blanks, and split or duplicate samples; trip blanks should be included when volatile organic compound analyses are being conducted. Laboratory blanks and spikes (i.e., blind blanks and spikes) should be submitted in three cases: for major sampling efforts (e.g., 20 or more samples), whenever a new laboratory is used, or when analytical problems are suspected. If possible, when laboratory blanks and spikes are used, the blanks and spikes should be prepared by a laboratory which is not performing the analyses, so that the blanks and spikes will be "blind" to the analytical laboratory.

Field quality control data are generally interpreted by either the quality control chemist or the project leader (i.e., principal report writer). The data are normally used to flag questionable data and, in most cases, interpretation has been on a case-by-case basis. Although a fairly large data base of field quality control data exists in association with the EPA contract laboratory program, no exhaustive analyses of these data have been undertaken and no standard procedures have been developed for acceptance or rejection of data based on field quality control samples. This is an area where additional research is needed.

SAMPLE HANDLING PROCEDURES

Sample handling procedures include sample collection, field storage, and shipment. During collection, care must be taken to ensure that the integrity of the sample is not compromised by either contamination from an extraneous source or loss of substances that are present in the sample such as highly volatile compounds. Sampling personnel should wear disposable gloves at all times during collection and mixing of samples to prevent contamination of the samples or sampling equipment and containers by substances on the samplers hands. Sampling equipment and containers including lids should be kept covered or closed until time for use. Care should be taken to avoid laying equipment and lids on the ground. Plastic sheeting or aluminum foil should be used to keep equipment off the ground.

A critical part of sample collection is mixing the sample to achieve homogeneity. Soil samples are often non-homogeneous. When samples are collected from "hot spots" or spill areas, extreme variation can result between laboratory replicates and/or field split samples if the samples are not properly mixed. Mixing in the field must be done with great care to avoid contamination of the sample from extraneous sources and to reduce the loss of volatile compounds. Mixing should be conducted in a deep pan to reduce the surface area of the sample exposed to the atmosphere and should be accomplished using a slow smooth motion to reduce aeration of the sample. Mixing should continue until the sample is visibly homogeneous. Mixing of samples by the laboratory under controlled conditions should be considered when possible to reduce the potential for sample contamination and loss of volatiles. Many production laboratories are not set up to provide such services; therefore, sample mixing must be accomplished in the field. Another instance which requires field mixing of samples is the submission of "blind" splits to the analytical laboratory.

After collection, lids should be placed on the samples immediately. Chemical preservation is not normally used for soil samples; however, samples should be kept cool (4°C) from time of collection until analysis to avoid degradation of compounds present in the sample. Head space in containers to be analyzed for highly volatile compounds (purgeable organic compounds) should be minimized to avoid losses resulting from volatilization. Although no published holding times are available for soil samples, water holding times should be observed where possible to reduce degradation and volatilization of components in the sample [5].

CONCLUSION

Quality assurance/quality control is an integral part of soil sampling activities just as it is in all other sampling activities. Sampling equipment, containers, and procedures must be carefully selected and used to minimize the possibility of affecting sample integrity. Proper equipment and container cleaning procedures must be used to avoid cross contamination. Quality control samples including blanks, spikes, and split/duplicates should be used to provide a measure of data quality.

REFERENCES

1. Barcelona, M. J., Jones, P. G., and Miller, R. A., A Guide to the Selection of Materials for Monitoring Well Construction and Groundwater Sampling, SWS Contract Report 327, Illinois State Water Survey, Department of Energy and Natural Resources, Champaign, Illinois, August 1983.

2. Boettner, E. A., Ball, G. L., Hollingsworth, Z., and Aquino, R., Project Summary-Organic and Organotin Compounds Leached from PVC and CPVC Pipe. EPA-600/S1-81-062. U.S. Environmental Protection Agency, Health Effects Research Laboratory, Cincinnati, Ohio, February 1982.

3. Curran, C. M. and Thomson, M. B., "Leading of Trace Organics into
    Waters from Five Common Plastics," Groundwater Review, National
    Water Well Association, Summer 1983.

4. McDaniels, W. H., Personal communication, Re: Use of Aluminum
    Pans During the Collection of Soil Samples, U.S. Environmental
    Protection Agency, Environmental Services Division, Analytical
    Support Branch, 1984.

5. U.S. Environmental Protection Agency, Draft Waters Surveillance
    Branch Standard Operating Procedures and Quality Assurance Manual,
    Surveillance and Analysis Division, Athens, Georgia, August 19,
    1980.

6. Lair, M. D., "Memo to Finger/Adams, et al., Re: Solvent Used to
    Clean Sampling Equipment," U.S. Environmental Protection Agency,
    Environmental Services Division, September 10, 1984.

7. Bruner, R. J., III, "Sampling Schemes and Site Selection,"
    Lecture: RCRA and Hazardous Waste Site Sampling Short Course
    Presented to States and Federal Agencies during 1983-85.

8. U.S. Environmental Protection Agency, Preparation of Soil Sampling
    Protocols: Techniques and Strategies, EPA 600/4-83-020, Environ-
    mental Monitoring Systems Laboratory, Las Vegas, Nevada, August
    1983.

9. U.S. Environmental Protection Agency, Test Procedures for Eval-
    uating Solid Waste, SW-846, Office of Solid Waste Management,
    Washington, D.C., 1982.

George T. Flatman

DESIGN OF SOIL SAMPLING PROGRAMS: STATISTICAL CONSIDERATIONS

REFERENCE: Flatman, G. T., "Design of Soil Sampling
Programs: Statistical Considerations," Quality
Control in Remedial Site Investigation: Hazardous and
Industrial Solid Waste Testing, Fifth Volume, ASTMSTP925,
C. L. Perket, Ed., American Society for Testing and
Materials, Philadelphia, 1986.

ABSTRACT: This paper identifies the critical problems of
soil sampling for pollution monitoring and explains the
solutions provided by new advances in spatial statistics
such as Kriging. The U.S. EPA at the Environmental Monitoring Systems Laboratory, Las Vegas, is applying new types
of spatial statistics to the chronic problems of pollution
plume monitoring and contouring. Firstly, the most common
soil sampling problem is the failure to recognize and utilize autocorrelation in space and/or time of the variable in
question. Such correlation variables require a systematic
rather than a random sampling design, and a spatial analysis by contouring rather than a t-test. This problem can
be solved by the use of Kriging. Secondly, chemical analyses of pollution samples often give many readings below
detection limit with a few relatively high readings indi-
cating a non-normal, truncated, positively skewed distribution. Estimates or tests assuming normal theory are
imprecise for such data, but nonparametric spatial statistics are more appropriate. Thirdly, data may be from
different monitoring techniques or different chemical
analyses, causing heteroprecision. Such data require an
analysis such as "soft-data" Kriging that accounts for the
precision of each datum in the analysis. Finally, if
remediation decisions are to be based on the data, the
probability of misjudging "clean areas to be dirty" or
"dirty areas to be clean" must be calculable. Probability
Kriging gives such probabilities. This paper discusses
these new forms of spatial analysis as they apply to the
recurring problems of pollution monitoring.

KEYWORDS: spatial statistics, Kriging, soil sampling,
sampling design, monitoring system design, geostatistics

Mr. Flatman is a Mathematical Statistician at United States
Environmental Protection Agency, Environmental Monitoring Systems
Laboratory-Las Vegas, Las Vegas, Nevada 89114.

43

INTRODUCTION

The growing number and complexity of toxic chemicals and hazardous waste sites call for more efficient sampling designs and more precise data analyses. In response to this need, the Environmental Monitoring Systems Laboratory--Las Vegas has initiated a Monitoring Statistics Project to develop appropriate field sampling designs and field data analyses. The disciplines of statistics and computer science are growing rapidly, producing more appropriate algorithms and more usable computer programs.

Due to these improvements, field sampling and analysis methods must be revised continually. The sampling design that was appropriate and used a few years ago may be second choice today. The sampling design and data analysis must agree with the variable's type, distribution, precision, and the monitoring program's goal. This paper will discuss the sampling and analysis of (1) spatially auto-correlated variables, (2) truncated skewed distributions, (3) hetero-precise measurement techniques, and (4) the probabilistic presentation of sampling information for the decision makers.

SAMPLING SPATIAL VARIABLES

In pollution monitoring, because physical laws govern the source, fate, and transport of a pollutant, the contamination concentration values over the sampling site are often spatially correlated. This is illustrated by the intuitive fact that samples closer together in location are also closer together in concentration values, and samples more distant in location are usually more divergent in concentration values. Variables that are spatially correlated should not be sampled, tested, or analyzed as if they were independent. The use of random variable methods on spatially correlated variables may give biased estimates, and tests with imprecise degrees of freedom and levels of confidence. The use of spatial variable methods on spatially correlated variables will use the spatial correlation structure to decrease the number of samples required and give regionally representative estimates. Spatial dependence is easily detected by plotting the concentration values on an area map. If the large concentrations are in definite regions and the small concentrations are in definite regions, there is spatial correlation.

Representative Sample

The representativeness of a random variable is often defined more with rhetoric than with mathematical equations. The trustworthiness of some random sampling programs has been discredited by asking whether the samples are representative of the population or if they are representative only of themselves. The representativeness of a spatial variable is defined by the "range of correlation" of the "spatial correlation structure" of the pollutant plume and can be measured by the "semivariogram" [1]. The "range of correlation" is the maximum distance between sampling locations at which the samples are still correlated, or the minimum distance between sampling sites at which samples are independent. Thus, the spatial representativeness of a sample is the geographic area defined by a radius centered at the sample site and of length equal to the range of correlation.

Semivariogram

The semivariogram is a geostatistical tool calculated from information about time or space relationships between sample observations. The semivariogram in Figure 1 graphically shows the relationships between observations in terms of their separation (lag). The x-axis represents the lag or distance between sample points, and the y-axis represents gamma or the "variance" of the difference in pollutant levels between pairs of samples at each lag.

Since both the distance and the difference between two samples of correlated variables taken at the same point are zero, the semivariogram curve theoretically passes through the origin. As the number of lags increases, the correlation between the pairs of samples decreases, and the variance of the difference in their values increases, resulting in a rising curve or semivariogram. As the number of lags between samples becomes sufficiently large, the sample values become independent of each other, the variance of their differences becomes

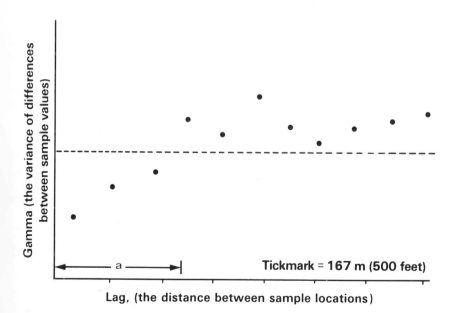

FIG. 1--A semivariogram of lead samples taken systematically on a 230m (750 foot) grid.

nearly constant, and the curve becomes horizontal. The distance along the x-axis where the semivariogram is rising represents the "range of correlation," the distance within which samples are related. This range and the semivariogram are important factors in determining the grid design for sampling, in contouring pollutant concentrations on a map, and in calculating the uncertainties in contouring [2]. Often the structure, and therefore the range of correlation, is different for different directions across a pollution plume or mineral deposit. This lack of equality in structure is called anisotropy and requires the computation of variograms in various directions. Directional variograms are often computed in 45° angular increments around the compass.

The best (minimum variance) sampling design for random variables is random; the minimum variance design for spatially correlated variables is systematic, i.e., a grid sampled at all its vertices. The grid may be made up of squares, triangles or hexagons [3] as indicated by the parameters of the semivariogram [4]. The side of the grid, i.e., distance between sampling locations, should be less than the range of correlation but more than one half that range for optimum sampling. The decrease of variance with decrease in grid size reaches a point of diminishing returns for information gained versus number of samples as the grid size approaches one half the range of correlation [4].

If the directional semivariograms have markedly different ranges of correlation (geometric anisotropy), then the side of the grid should be scaled by the ratio of the ranges of correlation of the directional semivariograms. That is, the grid square would become rectangular [5], the grid equilateral triangle would become isosceles or scalene, the grid regular hexagon would become irregular. The longer axis of the sampling grid should parallel the direction of the directional semivariogram with the longest range of correlation. Thus, the directional semivariograms give information for the sampling grid's size, shape, and orientation.

ANALYZING SPATIAL VARIABLES

When sampling or analyzing spatial variables, several areas or volumes must be chosen to fit the purpose of the investigation. For clarity, let the total area suspected of being polluted be called the "region in question." A sample has at least three important areas or volumes. The first is the input unit area for which the sample is representative if the grid side is determined from the range of correlation. In checking an output isomap, it is important to think of these input unit areas as a grid which is offset so that the sample location is the center rather than the vertex of the square. Secondly, the sample also has its own volume or geometry, called the support. For composite samples, multimedia samples, or geophysical readings, the support is an important factor in the analysis. Lastly, there is an output unit area, called a block, for which an estimate is calculated. The block is best determined by the purpose of the sampling

program. If the isomaps will be used to guide remediation, the block
might be determined by the volume to be moved in one working period.
A spatial statistical analysis estimates a mean and interpolation
error for each output unit area and gives a contour map of spatial
mean and a contour map of spatial interpolation error for the region
in question. The mean or variance of a population of random variables
is a single value, but the mean or variance of spatial variables is a
contour map of the region in question. In spatial analysis each
input sample is a concentration representing grid area or unit input
area, and each output estimate is a concentration representing a
point or unit output area. The Kriging analysis gives a mean and
variance for each output point or block. Joining the points or blocks
of equal concentration within the region in question produces the
contour map over the entire region. The output unit area may be dic-
tated by the cleanup strategy.

Isomaps or Contour Maps of Concentrations

Kriging gives contour maps of the estimated mean and standard
deviation per unit area as shown in Figures 2 and 3. In Figure 2,
note that the abscissa is distance west to east in feet and the
ordinate is distance south to north in feet. A smelter located in
the center marked by a circled "X" is enclosed in a nested set of
irregularly elliptical closed contours. The contours show a steep
gradient from the outside contour of 300 ppm to the innermost contour
of 2,500 ppm. If lead were deposited chiefly by automobiles, one
would expect the contour lines to nest along the major roads of the
area. This was not apparent.

However, the nest of the closed contour lines around the smelter
suggests that the plume was caused by the smelter. This contour map
was put on transparent Mylar and scaled to overlay the aerial photo
so that precise identification of the "dirty" and "clean" areas could
be made.

Figure 3 is a contour of the standard deviation or the error of
estimation. The Swiss cheese appearance (small circles around the
sample locations) shows that the contour map of lead concentrations
is more accurate near sampling locations. Spatial analysis by Kriging
quantifies this degree of accuracy. The contour map of errors shows
that the sampling grid gives multiplicative standard deviations less
than two in the sampled region, but for the large circular contours
outside the sampled region the multiplicative standard deviations
exceed two. The large circles around the outside show how quickly
precision falls off beyond the data locations. This standard deviation represents an interpolation error and an evaluation of the grid
spacing. Because of the skewness of the data and the skewness of
the estimates, any confidence interval assuming normality would be
inaccurate.

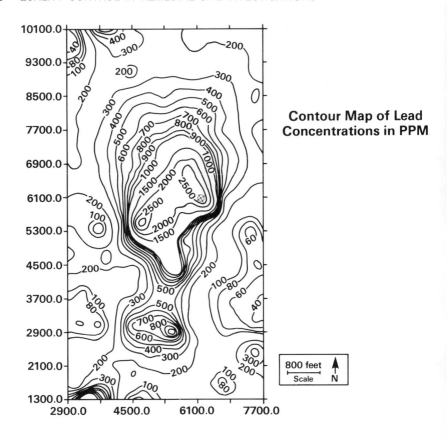

FIG. 2--Contour map of the lead concentrations in ppm around
the smelter.

ANALYZING TRUNCATED SKEWED DISTRIBUTIONS

    The monitoring specialist may find many sample values reported by
the analytical labs as below detection limits, especially if the pollutant of interest is a volatile or semivolatile. This situation may
be caused by an inadequate volume of the sample rather than the lack
of pollution of the area represented by the sample. The sample value
of "below detection limit" is not a value but a range of numbers, and
thus precludes arithmetic operation or summary statistics. This
arithmetic uselessness of "below detection limit" is more than a mathematical problem if the action level for the pollutant is of the same
magnitude as the detection limit. However, the detection limit problem is usually linked to a positively skewed distribution in pollution
monitoring. That is, in a set of sample values, there are many samples with low concentrations of pollution and a few samples with high

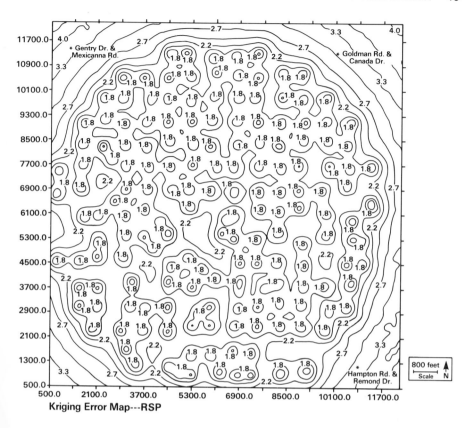

FIG. 3--Contour map of standard deviations of the lead
concentrations in ppm.

concentrations. The arithmetic mean is a misleading statistic of
data so distributed because it is overly influenced by extremely high
values.

Indicator Kriging and Probability Kriging

To more appropriately contour this frequent type of monitoring
data with many low or below-detection-limit values and a few very
high values, Dr. Andre Journel has developed, in cooperation with
United States Environmental Protection Agency's Environmental Monitoring Systems Laboratory at Las Vegas, a nonparametric probability
spatial analysis [6]. The problems of truncation (below detection
limit) and positive skewness (outliers) are minimized by the use of
indicator variables and rank order statistics [7]. These venerable
nonparametric tools are used to generate an empirical frequency function for each unit area [8]. Thus, the isopleths drawn on the isomap
are not pollution concentration but probabilities of that concentration exceeding or falling below some action level. Probability is a

convenient transformation for any applied data because it has meaning in the "real world." There is no need for an inverse transformation that reintroduces the distortion that necessitated the original transformation. The monitoring specialist or the remediation decision maker can understand or use the "probability of pollution level exceeding the action level" as easily as the "pollution level exceeding the action level." Since probability is calculated in terms of "greater than or less than" some value, it is an excellent tool for talking about "less than or greater than" the action level. Figure 4 shows the contour lines of the probability in percent that the lead concentration exceeds 1000 ppm around the smelter of Figures 2 and 3.

HETEROPRECISION OF DATA

Quality control or quality assurance shows that the concentration given by a chemical analysis is not actually a point value as often assumed for ease of calculation, but, in fact, the concentration is a continuum of values. If subgroups of the data were analyzed at different laboratories or by different methods, it would be probable that each subgroup would have a different standard deviation or range. These ranges are often noted in tables of the data by plus and minus signs followed by the standard deviation. The standard deviation is a measure of dispersion and is the square root of the variance. The standard deviation gives the confidence interval (i.e., the range within which the true value of the parameter being estimated is to be found with a given probability). A plus and minus one standard deviation interval gives a 68% probability of containing the true value.

This heterogeneity of precision becomes larger and more important to statistical analysis when different investigative techniques are used. In ground-water monitoring, the data might consist of a few water sample analyses which are relative point values and of many geophysical values which are "soft-data" or ranges within which the true value lies with some probability.

In mathematical purity, no "real world" measurement is a point value, but the concepts of "hard" or "soft-data" are relative. The well data are point data in relation to geophysical interval data. A second type of "soft-data" might be called "probabilistic-soft-data" and is the interval datum with the added information of the general shape of the distribution of the true value within the interval. The distributions are thought of in a simple classification, e.g., rectangular, isosceles triangular, or scalene triangular. The rectangular distribution shows each point in the interval is equally likely (uniform). The isosceles triangular distribution shows a symmetrical distribution with a strong central tendency (weakly normal). The scalene triangular distribution representing skewed distributions shows that values at one end of the interval are more likely. "Probabilistic-soft-data" is needed because many users assume isosceles triangular distributions (weakly normal) while many nonparametric algorithms assume rectangular distributions (uniform). In the inverse

Contour Map of the Probability in Percent of Finding the Value of 1,000 PPM or a Larger Value

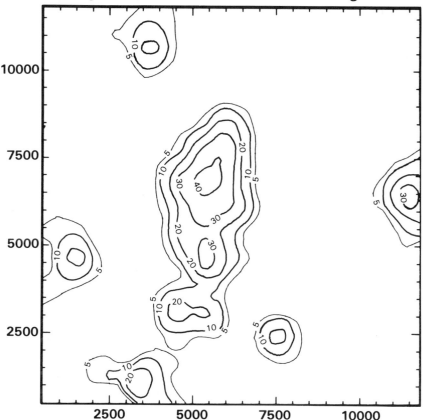

FIG. 4—Contour map of probabilities in percent that the lead concentration exceeds 1000 ppm.

modeling and interpretation of geophysical data, these types of distribution are suggested. Such mixtures of data need a "soft-data" Kriging algorithm that takes into account precision as well as distance between samples.

Soft-Data Probability Kriging

The indicator-probability algorithm can be extended to use "soft-input."[9] A datum may be a point value called a "hard-datum," an interval value called a "soft-datum," or an interval value with a frequency shape [10] called a "probabilistic-soft-datum." The "soft-datum" is used in the order statistics and the rank ordering as a trichotomous variable (less than, unknown, or greater than). The interval or interval-with-probability input data gives the spatial estimation algorithm the ability to weight each datum inversely to the variance of its precision as well as the variance of its distance from the point to be estimated. This means that among equally distant data locations, the most precise datum will be given the most weight in calculating the estimate. For example, in estimating a point equi-distant from well data and geophysical readings, the well data would have more influence on the answer than the geophysical data. Taking this second source of variation into the spatial estimation algorithm will model the monitoring problem more precisely and ought to give more precise estimates. This more complicated Kriging of the data is necessary for the statistical analysis of ground water where well data and geophysical readings or other multimedia data are taken.

DECISIONS AND ERRORS

Evaluating risk assessment or remediation for a pollution site is not a clear-cut decision. There are often data implying the site is "clean" and data implying the site is "dirty." Since monitoring data seldom unanimously show the site to be "clean" or "dirty," any judg-ments based on them are made at a confidence level with a probability of being right and a probability of being wrong. In a remediation decision, the party paying for the cleanup may ask, "How sure is the administrating agency that the area to be treated needs treating?" and the concerned citizens may ask, "How sure is the administrating agency that the area to be left untreated is truly safe to be un-treated?" Both of these questions are honest concerns and worthy of an answer. Rephrased in statistical terms, these questions ask, "What is the probability that the area to be treated has a pollution concen-tration less than the action level?" and, "What is the probability that the area to be left untreated has a pollution concentration ex-ceeding the action level?" Such probabilities should be calculated in any monitoring data analysis and made available to the decision makers and interested parties.

Contour Maps of the Probability of Misclassification

Figure 5 is a probability Kriging isomap that quantifies the probability of misclassifying a clean area as a "dirty" area. The outside isopleths labeled 1000 are the lead contours bounding the areas where the lead concentration exceeds 1000 ppm. For purpose of illustrating a false positive, assume 1000 ppm is the action level, and thus the area enclosed by this contour is judged "dirty" and will be cleaned up. The isopleths labeled 80, 70, and 60 are probability contours inside the area judged "dirty." They record the probability in percent that the lead concentration is less than 1000 ppm. That is the probability of cleaning up a clean area. These probabilities are low compared to the statistical textbook examples of confidence level (.95 or .99). There are at least two reasons for these differ-ences: firstly, the estimates are positively skewed; and secondly, monitoring is an uncontrolled experiment. The positive skewness of the estimates as well as the values of the input data illustrate the bias that would be introduced by assuming normality. Following normal theory and considering the estimates to be a weighted average, the

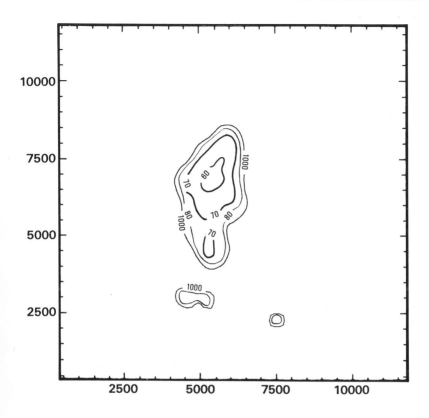

FIG. 5--Contour map of probabilities in percent that the lead
concentration is less than 1000 ppm in area estimated.

probability of being above the mean should be equal to the probability
of being below the mean, i.e., 50 percent. In reading the textbook
example, the experimental goal sounds misleadingly similar; the textbook is testing crop yield between plots while the monitoring program
is testing pollutant concentration between plume and background. The
difference is that the textbook experiment has the contributing fac-
ors controlled by the experiment station while the real world pollu-
ion site measurement experiment neither controls nor even measures
any of the contributing factors. More samples would improve these
robabilities, but time and sampling dollars are limited.

Figure 6 is a probability Kriging isomap that displays the prob-
bility of misclassifying a dirty area as "clean." The inside iso-
leths labeled 1000 ppm are the lead contours bounding the area where
he lead concentration exceeds 1000 ppm as in Figure 5. Again assume
or illustration that 1000 ppm is the action level so that the area
utside this contour is judged "clean" and will not be treated. The

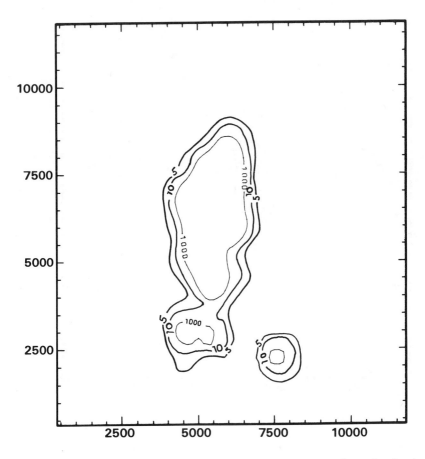

FIG. 6--Contour map of probabilities in percent that the lead
concentration exceeds 1000 ppm in area estimated to be less
than 1000 ppm or false negative).

outside isopleths labeled 5 and 10 are probability contours in the
area judged "clean." They record the probability in percent that the
lead concentration is higher than 1000 ppm. This is the probability
of not cleaning up a dirty area [11]. Note that these probabilities
of a false negative are lower than the probabilities in Figure 5 of
a false positive. The two are inversely related in regard to the
boundary value. If the boundary value were raised, the probability
of false positives (unnecessary cleanup) would be lessened, but with
an increase in the probability of a false negative (leaving a "dirty"
area). Thus the relative magnitude of the probabilities of the false
positive and negative should be inversely proportional to their costs
[12]. If the cost of a false positive (incremental cleanup of additional area) is less than the cost of a false negative (health risk

due to not cleaning an area), then the larger probability of false positive is acceptable.

In the complexity of the real world, the decisions of risk assessment and remediation are value judgments made from limited data. They cannot be mechanically made by a statistical algorithm, but probabilistic Kriging can give the decision makers geographic contours of any given pollution concentration (action level) and geographic contours of the probabilities of a false positive or negative. Such tools will be of help to the decision makers and to all parties concerned.

CONCLUSIONS

The decisions of environmental protection in regard to risk assessment and remediation are complex, but modern technologies are providing better decision tools. Chemistry is contributing more economical and more accurate analyses. Monitoring science is producing more economical and more interpretable remote sensing. Statistics also has a contribution to cut sampling costs and to improve precision. Many of the chronic problems of pollution monitoring can be addressed by probabilistic soft-data Kriging. The use of these statistical tools may necessitate the addition of a statistician to the team, but sampling costs may be cut and the data analysis outputs will be more precise and more easily understood.

REFERENCES

[1]   Flatman, G. T., "Geostatistical Strategy for Soil Sampling: The Survey and the Census," Environmental Monitoring and Assessment, Vol. 4, December 1984, pp. 335-349.

[2]   Clark, I., Practical Geostatistics, Applied Science Publishers Ltd., London, 1982.

[3]   Olea, R. A., Systematic Sampling of Spatial Functions, Series on Spatial Analysis No. 7, Kansas Geological Survey. University of Kansas, Lawrence, Kansas, 1984.

[4]   Yfantis, A. A., Flatman, G. T., and Behar, J. V., "Sampling of Spatial Second Order Stationary Random Processes," submitted to Journal of Mathematical Geology.

[5]   David, M., Geostatistical Ore Reserve Estimation, Elsevier Scientific Publishing Company, Amsterdam, Oxford, New York, 1977.

[6]   Journel, A. G., "Indicator Approach to Toxic Chemical Sites," EPA report, Project No. CR-811235-020, 1984.

[7]   Journel, A. G., "Non-Parametric Estimation of Spatial Distribu-
      tions," Journal of Mathematical Geology, Vol. 15, No. 3, July 1983,
      pp. 445-468.

[8]   Journel, A. G., "The Place of Non-Parametric Geostatistics,"
      in Geostatistics for Natural Resources Charaterization, ed.
      G. Verly, et al., pub. Reidel, Dordrecht, Holland, Part 1,
      1984, pp. 307-335.

[9]   Kulkarni, R., "Bayesian Kriging in Geotechnical Problems," in
      Geostatistics for Natural Resources Characterization, ed. Verly,
      et al., publ. Reidel, Holland, Vol. 2, 1984, pp. 775-7.

[10] Kostov, C., "Constrained Interpolation in Geostatistical Applica-
      tions," Thesis, Master of Science, Applied Earth Sciences Depart-
      ment, Stanford University, 1985.

[11] Isaaks, E. H., "Risk Qualified Mappings for Hazardous Wastes
      Sites. A Case Study in Non-Parametric Geostatistics," Thesis,
      Master of Science, Applied Earth Sciences Department, Stanford
      University, 1984.

[12] Flatman, G. T., and Mullins, J. W., "The Alpha and Beta of
      Chemometrics," American Chemical Society Annual Meeting,
      Philadelphia, 1984; available as American Chemical Society
      Symposium Series 292, Chapter 14.

Thomas H. Starks, Kenneth W. Brown, Nancy J. Fisher

PRELIMINARY MONITORING DESIGN FOR METAL POLLUTION IN
PALMERTON, PENNSYLVANIA

REFERENCE: Starks, T. H., Brown, K. W., and
Fisher, N. J., "Preliminary Monitoring Design for
Metal Pollution in Palmerton, Pennsylvania,"
Quality Control in Remedial Site Investigation:
Hazardous and Industrial Solid Waste Testing,
Fifth Volume, ASTM STP 925, C. L. Perket, Ed.,
American Society of Testing and Materials,
Philadelphia, 1986.

ABSTRACT: Large concentrations of the metal soil
pollutants, cadmium, copper, lead, and zinc, have
been detected in and around Palmerton, Pennsylvania.
Before a cleanup of the area can be undertaken, it
will be necessary to map the spatial distribution of
the concentrations of the pollutants over the area.
Then boundaries can be determined for the regions
that are above and below action levels. The estimation of the spatial distributions of the metal
concentrations, with a prespecified level of precision, requires a definitive soil-sampling survey.
The intensity of sampling (i.e., distance between
sample points) needed to attain the prespecified
level of precision will depend on the nonsampling
error variance and the spatial structure of the
concentration measurements. To determine the nonsampling error variance and the spatial structure,
a preliminary soil-sampling study is required.
This paper discusses the planning of such a preliminary study.

KEYWORDS: metal pollutants, smelter pollution,
soil sampling, spatial structure

Dr. Starks is a senior statistician at the Environmental
Research Center of the University of Nevada, Las Vegas, NV
89154; Mr. Brown is a botanist with the Environmental
Monitoring Systems Laboratory, USEPA, P.O. Box 15027, Las Vegas,
NV 89114; Ms. Fisher is a senior computer analyst at Computer
Sciences Corporation, Las Vegas, NV 89114.

INTRODUCTION

Atmospheric release of zinc, cadmium, copper, and lead by the New Jersey Zinc Company's Palmerton, Pennsylvania smelter was suspected of causing widespread environmental contamination [1]. Therefore, during the early 1980's, soil samples from selected sites in the vicinity of the smelters were collected by teams from the Pennsylvania State University (PSU) and from the United States Environmental Protection Agency's Health Effects Research Laboratory (HERL) in North Carolina.

To identify contaminant distribution, soils collected for the PSU study were obtained within a 25 km (15 mi) radius of the smelter. The sampling sites and the methods employed for this study, as previously reported by Washington [2], involved collecting soil samples from cultivated fields by using soil augers. The number of individual cores taken from each field and the core depth was determined by the size of the field being sampled and the depth of the Ap (plow depth) horizon. After collection, the cores were placed in a single container, dried, sieved, and mixed. An aliquot was taken from the mixed portion, acid digested, and then analyzed by atomic absorption spectroscopy (AAS) for total zinc, cadmium, lead, nickel, copper and iron.

The objective of the HERL study was to identify the relationship between heavy metal levels in biological tissue, i.e., human blood, hair, and urine, with metal levels found in air, soils, dust, and drinking water. Because of the diversity of this study, soils were collected from residential play areas, schools, and in the immediate vicinity of high volume air samplers. Only surface soils (10-15 cm) (4-6") were collected as exposure to subsurface soil contamination was considered unlikely. Similar to the PSU study, most of the HERL study soils were collected within a 25 km radius of the smelter in Palmerton. After collection, as previously described by Handy et al. [3], the soils were dried, sieved, and mixed. The soils were acid digested and analyzed by AAS for arsenic, cadmium, copper, manganese, lead, and zinc.

The results of these studies, in addition to studies by Buchauer [1], Jordan [4], Beyer et al. [5], and summarized by the NUS Corporation in 1984 [6], showed that elevated concentrations of zinc, cadmium, lead, and copper were found in vegetation and surface soils up to 10 miles west and 16 miles east of the Palmerton smelter. The soil values decreased as a function of distance from the smelter; however, values exceeding 130,000, 1700, 2200, and 1500 ppm for zinc, cadmium, lead, and copper respectively were found within 2 miles of the smelter.

Because of these high soil levels and levels of zinc and cadmium found in ground water, surface water, and biota collected near the smelter [7], the Palmerton zinc smelter site was added to EPA's Superfund National Priority List for remedial investigation.

In July 1984, the EPA's Environmental Monitoring Systems
Laboratory in Las Vegas (EMSL-LV) received a request from EPA's
Region 3 office in Philadelphia to assist in the development of
a soil monitoring program. The EMSL-LV was requested to:

o Design a soil-sampling network (i.e., grid size,
   spacing, and number of samples).

o Recommend soil-sampling techniques (i.e., sampling
   equipment, quantity of samples).

o Use geostatistics to provide soil concentration
   isopleths for cadmium, lead, zinc, and copper.

o Provide onsite sampling audits.

This report presents the rationale and methods for development of a sampling strategy to generate court-worthy data. Geostatistical estimation procedures [8] will be applied to the
data to obtain metal-concentration isopleths. The isopleths
will delineate the regions needing remedial action in the
vicinity of the Palmerton zinc smelter. The sampling methods
and analytical procedures have been previously reported [9,10].

SITE DESCRIPTION

    The Palmerton zinc complex occupies approximately 267 acres
near the city of Palmerton, Pennsylvania. This industrial
complex consisting of two separate plants is located west and
east of the city. The west plant is located on the northern
bank of the Lehigh River, at its confluence with Aquashicola
Creek; the east plant including the slag pile is located on the
southern bank of the Aquashicola Creek [6].

    Smelting operations began in 1898 at the western plant
where relatively pure zinc silicate was processed. Following
the construction of the eastern plant in 1915, zinc sulfide
containing approximately 55% zinc, 31% sulfur, 0.15% cadmium,
0.30% lead and 0.40% copper was processed. During the proces-
sing and smelting operations, as reported by NUS [6] the oxides
of sulfur, zinc, cadmium, and lead were incompletely recovered.
Daily metal emissions were estimated to be 5,900 to 9,000 kg
(13,000 to 19,800 lbs) per day of zinc and from 70 to 90 kg
(154 to 198 lbs) per day of cadmium. In 1950, the facility was
equipped with more efficient pollution controls and the cadmium
emissions dropped to about 23 kg (50 lbs) per day.

    The geological makeup of this area consists of a series of
deep, narrow valleys. The nearby ridges and valleys are under-
lain by thin, nearly vertical shale, siltstone, sandstone, and
limestone beds. Palmerton lies in a valley in which Chestnut
Ridge delineates the valley to the north, Aquashicola Creek
runs the length of the valley, and Blue Mountain borders the
valley on the south. The Lehigh Gap cuts through Blue Mountain
just east of the west plant and south of Palmerton [2].

This intensely folded and faulted area lies on the southern limb of the Wein Mountain Syncline. The syncline is east-northeast trending with the axis lying approximately 3 km (2 mi) north of Palmerton [6].

Glacial deposits and naturally occurring soils are present. The glacial unconsolidated deposits consist of brown to yellow brown sand, gravel, and cobbles. These deposits are horizontally stratified, ranging up to 18 m (60') thick in places. In some areas the glacial deposits are poorly sorted, stratified sandy gravels with interbedded red clay, ranging up to 24 m (80') in thickness.

The naturally occurring soils belong to the Klinesville, Holly, and Leck Kill series. Klinesville soils consist of shallow, 15 to 46 cm (6-18"), well drained reddish soils formed from sandstone, siltstone and shale. Holly soils are moderately deep, poorly drained, alluvial soils found near streams. The Leck Kill soils, located primarily near the slag pile, are moderately deep, well drained, acidic, brown residual soils formed from sandstone, siltstone, and shale (NUS, 1984).

Local wind and precipitation patterns are influenced to some extent by the topography of the surrounding area. As shown in Fig. 1, winds occur primarily from the northeast and southwest directions. (It should be noted that the N in Fig. 1 is magnetic north and the wind rose plot must be rotated 10° clockwise to agree with polar north.) Washington [2] reported that the winds from the north occur primarily during the winter months. Winds from the south and west are believed to greatly influence the distribution of stack emissions by flowing through the Lehigh Gap and continuing in a northeast direction up the Aquashicola Creek valley.

Meteorological data from the New Jersey Zinc Company's weather station in Palmerton has reported an average yearly temperature of 12°C (53.4°F), a minimum temperature of −25°C (−13°F), and a maximum temperature of 40.6°C (105°F). Yearly precipitation averages nearly 1.1 m (43") per year.

MONITORING DESIGN

Historical information and data from recently conducted soil sampling studies were collected and evaluated. This was done to determine if concentration isopleths could be provided at reasonable levels of precision without additional sampling, or, if this was not possible, to determine whether the data gave enough information about the spatial structure of the metal concentrations to determine the density of sampling required for a definitive survey. The results of this evaluation showed the data, collected for other objectives, to be inappropriate and inadequate for either of the above purposes. Therefore, a preliminary survey had to be developed that would provide information on the spatial structure (i.e., spatial

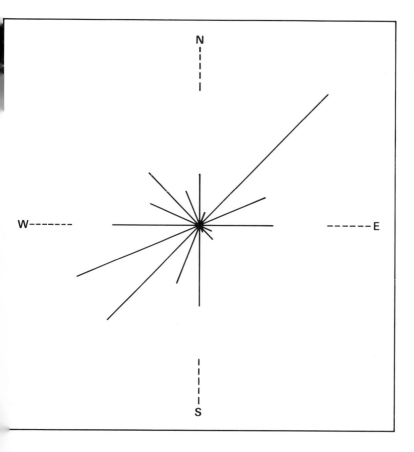

FIG. 1--Palmerton Wind Rose 1978-1979 data.

orrelation and order of drift) and extent of the concentra-
ions of the metal pollutants in the soil. The information
btained from this preliminary study will be used in the plan-
ing and development of a definitive monitoring study. In
ddition, the preliminary survey will provide data that will
e used along with the definitive study results in mapping the
istribution of the pollutant concentrations in the Palmerton
rea.

One suggested procedure for sampling to determine spatial
ructure was to take equally spaced samples along four tran-
ects: N-S, E-W, NE-SW, and NW-SE. However, it was deter-
ned that this would be wasteful of sample points in that each
mple point would in general contribute to the estimation of
atial correlation in only one direction. On the other hand,
samples were on a square grid, another common approach, each
int could be used in the estimation of spatial correlations

in several directions. However, if the grid distance is sufficiently small to estimate short range correlations, there may
be too few, if any, pairs of points available to estimate long
range correlations and the extent of pollution. The solution
was a compromise in which both the grid and the transects were
employed. The grid portion allows precise estimation of short
range correlations and the transects permit estimation of
extent and long range correlations.

To determine sampling intensity (distance between points),
information from a previous soil-lead monitoring study conducted in Dallas, Texas [11] was utilized. The Dallas lead
study found a 366 m (1200') range of influence for lead measurements in the areas near the lead smelters. There are
obvious differences between Palmerton and Dallas with regard to
topography and variability of soil types. In addition, the
Palmerton study involves three additional metal pollutants.
Nevertheless, lacking a better model for spatial variability,
it was decided that the spacing should be sufficiently small to
allow estimation of spatial correlation at several distances
less than 366 m. A good estimate of the spatial correlation
model, especially for short distance, is important in that it
determines the interpolation between points in the data analysis, and estimation of standard errors of the interpolations.
Using the results of the Dallas soil lead data and best judge-
ment a grid spacing of 122 m (400') was selected. The spacing
allows estimation of spatial correlations for lags of 122, 244,
and (on the diagonal) 172 m (400', 800', and 566) where spatial
correlations should be positive assuming a range of 366 m. The
use of spacings less than 122 m would reduce the areal coverage
as shown on Fig. 2, of the preliminary survey or increase the
number of points required. Hence, no additional sampling of
the preliminary survey area will be required in the definitive
study. On the other hand, if sample point spacing considerably
larger than 122 m were employed in the preliminary study, the
data might not be sufficient to provide estimates of spatial
correlation at lags where it is significantly positive. Then
the estimation of the spatial structure model would become
tenuous or impossible. For example, if the true model for
spatial correlation can be represented by a spherical variogram
model with range 366 m, then the correlation between points 244
m apart is only 0.15; while at 122 m, it is 0.52.

In many geostatistical studies, the nugget effect a short
range spatial correlation must be determined by extrapolating
the experimental semivariogram results of the first few lags
back to the zero lag. However, in environmental soil sampling,
a procedure that has been successfully used involves the collection of duplicate samples (within 0.5 m) at 5% of the sample
points, and splitting a total of 5% of the other samples
collected. The duplicates give information on short-range
variability (true nugget effect) and the splits give information on the combined subsampling and analytical-error variance.
In models of spatial structure, these two variances are usually
added and the total is called the nugget effect. When using the
least squares estimation technique [8], kriging, they should

be treated separately because they effect the kriging variance
in different ways. Hence, the information from the duplicates
and splits is extremely useful in determining the model for
spatial structure, especially in the important short lag region.

After the sampling pattern was agreed upon, a positioning
of the pattern on the map of the Palmerton area was performed
(Fig. 2). The placement of the pattern was based on survey
constraints and hypotheses about the pollution dispersion
formed from information on wind roses and local topography.
The pattern was centered in Palmerton and rotated so that the
transects would follow the river valleys and run parallel with
the principal wind rose directions. Also, since the large
smelter property was not to be sampled, the grid and transects
were oriented so as to miss the property as much as possible.

The purpose of the transects was to obtain information on
the areal extent of the metal pollution. The lengths of the
transects were determined by wind rose data, topography, and
land use. The transects were longer in the principal wind directions and/or where they extended into farm or residential
areas. They were shorter in minor wind directions and/or where
transects extended into forested areas. The greatest distance
from the center of the grid to the end of a transect was 8.5 km
(5.3 mi). Information from the transects should provide good
estimates of the rates of attenuation of pollution in the transect directions. Except near the grid, Fig. 2, the spacing of
sample points on the transects is 366 m or 345 m (1131') depending on whether the transect direction is parallel to an edge of
the grid squares or parallel to a diagonal of the squares.

Costs for chemical analysis being higher than sample collection costs led to the decision to archive soil samples taken
at 30 points obtained on an "every-other-one" basis near the
ends of the longest transects. These samples will be analyzed
only if statistical analysis of the other data indicates a need
for additional information. Soil samples will be taken at a
total of 210 points (85 on the grid and 125 on the transects)
but only 180 will be submitted for chemical analysis in the
preliminary study.

The numbers of pairs of points at various lags and directions, provided by the 180 points to be sampled and analyzed in
the preliminary study, are given in Table 1. The point sources
of the metals indicate that the model for the spatial struc-
tures of their concentrations will include drift or trend. The
wind rose pattern and the topography of the area suggest that
the model will be anisotropic, i.e., the spatial structure varies
with direction. Hence, simple semivariogram models for spatial
structure are not likely to suffice. For this situation, an
intrinsic random function model will be needed. Once such a
model is fit to the data from the preliminary study, it will be
possible to predict the maximum estimation standard error in
kriging estimates at points for any particular grid size that
might be employed in the definitive study. Therefore, once the
precision desired for the estimation of metal concentration is

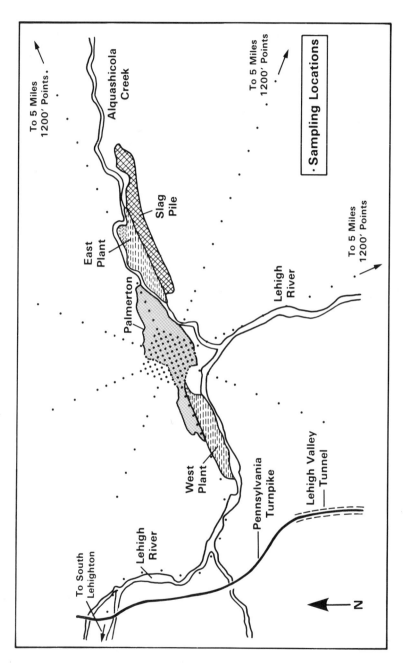

FIG. 2—Sample pattern for the initial Palmerton Survey (1" = 4250').

specified, the grid size for the final study can be determined.
The combined data from the preliminary and definitive studies
will then allow the kriging of estimates at points or on
blocks from which isopleth may be drawn to distinguish areas
above action levels from those that are below.

TABLE 1—Number of pairs in initial survey design for
estimation of experimental semivariogram.

Nominal	Direction		Nominal	Direction	
Lag[a]	E-W	N-S	Lag	NW-SE	NE-SW
122	85	78	173	82	82
244	79	71	345	74	79
366	72	70	517	58	58
488	64	57	690	59	55
732	61	54	1035	27	31
975	36	29	1379	25	21
1463	30	24			

[a]Lags in meters.

REFERENCES

[1] Buchauer, M. J., Contamination of Soil and Vegetation Near
     a Zinc Smelter by Zinc, Cadmium, Copper, and Lead,
     Environmental Science Technology, 7, 1973, pp. 131-135.
[2] Washington, D., Metal Concentrations in Soils and Ground
     Water in the Palmerton Area, Unpublished Thesis, The
     Pennsylvania State University, University Park, Pennsylvania,
     1984.
[3] Handy, R. W., Harris, S. H., Hartwell, T. D., and Williams,
     S. R., Epidemiologic Study Conducted in Populations Living
     Around Non-Ferrous Smelters, Volume 1 RTI/1372/00 Health
     Effects Research Laboratory, U.S. Environmental Protection
     Agency, Research Triangle Park, North Carolina, 1981.
[4] Jordan, M. J., Effects of Zinc Smelter Emissions and Fire
     on a Chestnut-Oak Woodland, Ecology 56, 1975, pp. 78-91.
[5] Beyer, W. N., Miller, G. W. and Cromartie, E. J., Con-
     tamination of the O_2 Soil Horizon by Zinc Smelting and Its
     Effect on Woodlouse Survival, Journal of Environmental
     Quality, 13, 1984, pp. 247-251.
[6] NUS Corporation, Work Plan - Remedial Investigation/
     Feasibility Study of Alternatives - Palmerton Zinc Site -
     Palmerton, Pennsylvania, EPA Work Assignment 63-3L26,
     Pittsburgh, Pennsylvania, 1984.
[7] NEIC, Evaluation of Runoff and Discharges from New Jersey
     Zinc Company, Palmerton, Pennsylvania, National Enforce-
     ment Investigations Center, U.S. Environmental Protection
     Agency, Denver, Colorado, 1979.
[8] Journel, A. G., and Huijbregts, Ch. J., Mining Geostatis-
     tics, Academic Press, New York, New York, 1978, 600 pp.

[9] U.S. Environmental Protection Agency, Palmerton Zinc NPL Site Investigation Soil Sampling Protocol, Environmental Monitoring Systems Laboratory, Las Vegas, Nevada, 1984.

[10] U.S. Environmental Protection Agency, User's Guide to the Contract Laboratory Program, Office of Emergency and Remedial Response, Washington, D.C., 1984a.

[11] Brown, K. W., Beckert, W. F., Black, S. C., Flatman, G. T., Mullins, J. W., Richitt, E. P., Simon, S. J., Documentation of EMSL-LV Contribution to Dallas Lead Study, EPA 600/4-84-012, U.S. Environmental Protection Agency, Las Vegas, Nevada, 1984.

Robert H. Laidlaw

DOCUMENT CONTROL AND CHAIN-OF-CUSTODY CONSIDERATIONS FOR THE NATIONAL
CONTRACT LABORATORY PROGRAM

REFERENCE: Laidlaw, R. H., "Document Control and Chain-of-Custody Considerations for the National Contract Laboratory Program," Quality Control in Remedial Site Investigations: Hazardous and Industrial Solid Waste Testing, Fifth Volume, ASTM STP 925, C. L. Perket, Ed., American Society for Testing and Materials, Philadelphia, 1986.

ABSTRACT: The Environmental Protection Agency (EPA) Contract Laboratory Program (CLP) provides analytical services for investigations in support of the Superfund program. The work effort includes data generation used for preparation of enforcement cases. EPA has specified procedures for evidence handling that are used to enhance admissibility of data and to establish analytical integrity. The CLP has stringent requirements for document control and chain-of-custody and has established a system of audits to assure compliance with procedure.

KEYWORDS: enforcement, Superfund, evidence, document control, audits

INTRODUCTION

The purpose of this presentation is to acquaint you with procedures that have been incorporated into the National Contract Laboratory Program (CLP) to satisfy Agency enforcement needs for investigations conducted under the Comprehensive Environmental Response, Compensation and Liability Act (CERCLA). Data generated from samples collected at uncontrolled hazardous waste sites have enforcement potential and the quality of the analytical results and the supporting documentation must be sufficient to support EPA litigation. The National Enforcement Investigations Center (NEIC) in Denver has established policies and procedures for evidence handling in contract laboratories and has worked with CLP management to require adherence to these procedures in the laboratory contracts.

The Comprehensive Environmental Response, Compensation and Liability Act of 1980, also known as Superfund, has enforcement provisions that

Robert H. Laidlaw is an Environmental Scientist at the Environmental Protection Agency, National Enforcement Investigations Center, Denver Federal Center, Denver, Colorado 80225.

rely in part on analytical data. The Act, under Section 104, authorizes the Government to respond to actual or threatened releases into the environment of any pollutant or contaminant which may present an imminent and substantial danger to the public health or welfare. In accordance with the National Contingency Plan, the Government may remove, or arrange for removal or may remediate the problem unless it is determined that responsible parties will take the necessary action. When the Government is authorized to act pursuant to Section 104, it may undertake investigations, monitoring, surveys, and other information gathering activities needed to identify the existence and extent of a release or threat of a release. Samples collected during this process are often analyzed by the CLP and the results can be expected to be used in EPA enforcement actions.

An imminent and substantial endangerment to the public health or welfare or the environment may bring action under Section 106 to secure relief through the court system. CLP data is used to support Section 106 enforcement actions. Under Section 107, responsible parties involved in the release or threatened release of hazardous substances which causes response costs to be incurred by the Government shall be liable for (1) all costs of removal or remedial action incurred by the Federal or State Government or by other persons operating in a manner consistent with the National Contingency Plan, and (2) damages for injury to, destruction of, or loss of natural resources, including costs of assessment. CLP analytical costs associated with CERCLA investigations can be recoverable under Section 107.

The significance of these enforcement activities is that CLP data and documents are evidentiary materials and, as such, must be able to withstand legal scrutiny. The analytical results and the paper trail must be both technically sound and legally defensible. CLP laboratories provide the same function for EPA Superfund investigations as any other forensic laboratory involved in law enforcement.

Commercial laboratories providing environmental testing services have traditionally been primarily concerned with reporting values and are less experienced in dealing with enforcement concerns. Laboratorie new to the CLP generally experience a learning curve in grasping the evidence handling perspective. Failure to adhere to the chain-of-custody and document control procedures can seriously damage EPA enforcement efforts.

CERCLA cases often involve large expenditures for investigation and remediation of the hazardous waste problem. The regulated community has a lot at stake in these investigations and corporations are tak ing these cases very seriously. They are becoming increasingly willing to challenge EPA findings and assessments of liability. There are numerous consultants available to the private sector that are extremely familiar with EPA procedures and case preparation strategy, and investigative and analytical procedures are reviewed with increasing scrutin EPA must have adherence to good evidence handling procedures by field and laboratory contractors to succeed in settlement or enforcement of Superfund cases.

CLP REQUIREMENTS FOR CHAIN-OF-CUSTODY AND DOCUMENT CONTROL

The chain-of-custody process is initiated in the field at the moment of sample collection. A chain-of-custody record (Figure 1) is used to list all samples, the date and time of collection, person(s) collecting them, and location identification. Each transfer of custody, including shipment to the laboratories is documented on the custody record.

CLP laboratories receive the secured container, inspect the contents, verify that all samples listed on the record are accounted for, and verify that the container hasn't been tampered with. The custodian then signs for receipt of custody.

Laboratory contracts contain a chain-of-custody and document control exhibit specifying the policies and procedures that must be in place to analyze samples for the CLP program. The contractor must develop written standard operating procedures (SOPs) for receipt of samples, maintenance of custody, tracking the analysis of samples and assembly of completed data. The SOP's must be a stepwise description describing how the laboratory will address each requirement. Guidance for preparation of the procedures is presented in the invitation for bid and generic models are available from the NEIC. The SOP's shall provide mechanisms and documentation to meet each of the following specifications:

° The contractor shall have a designated sample custodian responsible for receipt of samples.

° The contractor shall have written SOP's for receiving and logging in of the samples. The procedures shall include documentation of the sample condition, maintenance of custody and sample security, and documentation of verification of sample tag information against custody records.

° The contractor shall have written SOP's for maintenance of the security of the samples after log-in and shall demonstrate security of the sample storage and laboratory areas.

° The contractor shall have written SOP's for tracking the work performed on any particular sample. The tracking system shall include standard data logging formats, logbook entry procedures, and a means of controlling logbook pages, computer printouts, chromatograph tracings and other written or printed documents relevant to the samples. Logbooks, printed forms or other written documentation must be available to describe the work performed in each of the following areas:

   1. Sample receipt
   2. Sample extraction/preparation
   3. Sample analysis
   4. Data reduction
   5. Data reporting

° The contractor shall have written SOP's for organization and assembly of all documents relating to each EPA case. Docu-

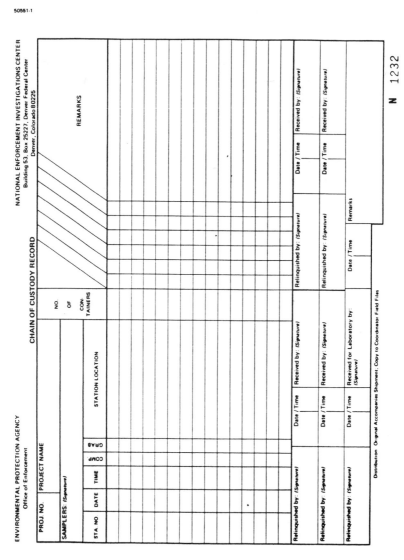

FIG. 1 – Chain-of-custody record

ments shall be filed on a case specific basis. The procedures must ensure that all documents including logbook pages, sample tracking records, measurement readout records, computer printouts, raw data summaries, correspondence, and any other written documents having reference to the case are compiled in one location for submission to EPA. The system must include a document numbering and inventory procedure.

Document control and chain-of-custody records include but are not limited to: sample tags, custody records, sample tracking records, analysts logbook pages, bench sheets, measurement readout records, extraction and analysis chronicles, computer printouts, raw data summaries, instrument logbook pages, correspondence and a document inventory. This group of documents is defined as the document control and chain-of-custody package and is an item required to be delivered to EPA. This group of documents comprises evidence that is needed to verify integrity of the sample analyses and to identify all potential laboratory witnesses. These records might well be used as trial exhibits and to prepare analysts for testimony.

EVIDENCE AUDITING

Reviews to assure compliance with the evidence handling requirements are conducted in the form of evidence audits. These audits are conducted prior to a contract award and also throughout the period of performance. The objective of the audit is to determine that the laboratory has the proper procedures in place and is following them.

Preaward audits are scheduled by the Environmental Monitoring Systems Laboratory in Las Vegas once the performance evaluation sample and documentation requirements have been satisfied. The preaward inspection is an onsite evaluation to determine that technical capability and quality assurance and quality control are in place and also to determine that evidentiary procedures are satisfactory. EMSL-LV has the technical evaluation responsibility and the National Enforcement Investigations Center has the enforcement role.

The contract laboratory preaward evidence audit consists of a review of standard operating procedures, an inspection of the laboratory facility, and observations regarding sample handling, security, and documentation practices. The laboratory must have the necessary SOP's, the facility must be secure, and there must be a document control system in place. A written report is prepared for the EPA Project Officer and Program Manager listing deficiencies that must be corrected prior to sample analysis. The report is submitted by the NEIC within ten days of the audit, and followup audits are conducted to ensure that deficiencies are corrected.

Laboratory evidence audits conducted during the period of performance of the contract are conducted by a Contract Evidence Audit Team that reports to the NEIC. The audit is a team effort with EMSL-LV which serves as the Team Leader. The Evidence Audit Team reviews the preaward report, the type of contract awarded, and the requirements of the particular contract including special analytical services.

The audit is initiated with a briefing with laboratory management and project leaders to discuss the purpose and scope of the audit. A walk-through inspection is used to observe procedures and documentation for sample receipt, sample storage, sample tracking, and case file assembly. The Contract Evidence Audit Team (CEAT) fills out a laboratory audit checklist provided by the NEIC and case files are reviewed to determine adherence to contract requirements. A verbal debriefing is held at the conclusion of the audit to inform laboratory managers of the findings and to discuss deficiencies that will be reported to the EPA Project Officer.

The CEAT prepares a report which is submitted to NEIC within ten days of the audit. The report contains a statement from the Evidence Audit Team certified public accountant that the laboratory either (1) met or exceeded evidence audit requirements, (2) failed to meet the requirements, or (3) met the requirements but had one or more deficiencies needing corrective action.

NEIC reviews the laboratory reports and transmits them to the Program Manager, Project Officer, and Deputy Project Officer. Direction to correct deficiencies is given by the Project Officer. Particular attention is given to repeated deficiencies and action is taken if laboratories fail to comply.

A second audit function involves case files for analytical work that has been completed. Contract laboratories initially deliver data packages to the EPA Client Office that contain the analytical results and the accompanying quality control and quality assurance information. The CLP administration, in January 1983, initiated an additional delivery specification that routes the remaining supporting documentation to the NEIC Evidence Audit Team for review prior to shipment to the client.

EPA enforcement may not occur for a period of years after the analyses are completed and the records that may be required for discovery, settlement negotiations, and witness preparation for trial need to be in EPA possession. Some contracts are completed or laboratories are no longer in business by the time the Agency needs to produce records and we must be assured that the documents are available and reviewed prior to enforcement activity.

The procedure calls for shipment of all supporting documentation to the NEIC between 90 and 120 days after submission of the original data. The CEAT receives the files and audits them for adherence to chain-of-custody and document control procedures. Deficiencies in the packages are reported to NEIC in the form of a discrepancy report which is forwarded to the EPA Project Officer and Deputy for corrective action. Audit of the completed case files provides information for the onsite evaluations and laboratory deficiencies are discussed during the audit.

CONCLUSIONS

Document control and chain-of-custody procedures are necessary components of the Contract Laboratory Program. The data uses include a high potential for enforcement action and may also have a significant effect on the Agency's ability to get voluntary compliance from the regulated community and our ability to settle cases on terms favorable to the Agency. Commercial laboratories might not provide these services on their own and specifications must continue to be required in laboratory contracts. Evidence audits are necessary to ensure continued compliance with the requirements and onsite evaluations must be scheduled on a regular routine basis.

Experience over the last 5 years in the CLP has demonstrated that approximately 12 percent of the laboratories meet evidence handling requirements without exceptions. Seventy nine percent met basic requirements but have one or more minor deficiencies needing correction, and nine percent have failed to meet requirements and have required administrative actions to resolve the problem before additional samples are analyzed.

The evidence audit program has strengthened evidence handling and documentation practices in the CLP and has enabled a wide variety of users of the data, including enforcement, to have access to well-organized, inventoried and controlled case files. The information generated by the CLP laboratories is transferred in an orderly and timely fashion and is complete and easily retrievable. The requirements for chain-of-custody and document control are stringent and demanding but they have satisfied evidentiary concerns and are instrumental in establishing admissibility of evidence.

Stanley P. Kovell

THE ROLE OF THE ENVIRONMENTAL PROTECTION AGENCY'S CONTRACT LABORATORY PROGRAM IN ANALYTICAL METHODS EVALUATION AND VALIDATION

REFERENCE: Kovell, S.P., "The Role of the Environmental Protection Agency's Contract Laboratory Program in Analytical Methods Evaluation and Validation," Quality Control in Remedial Site Investigation: Hazardous and Industrial Solid Waste Testing, Fifth Volume, ASTM STP 925, C.L. Perket, Ed., American Society for Testing and Materials, 1986.

ABSTRACT: The Environmental Protection Agency's Contract Laboratory Program (CLP) offers a unique opportunity to evaluate and validate new analytical methodologies under real-world conditions, utilizing the very laboratory community that will ultimately be using the methods. As a result of the established infrastructure and the laboratory resources available, the CLP has a viable system to test new methods with consistency and discipline. The CLP has established the Routine Analytical Services (RAS) and Special Analytical Services (SAS) programs which are both instrumental in methods testing and validation. The RAS program generates a large laboratory performance data base which is directly applicable to the evaluation of new analytical methods. The potential for utilizing the CLP for evaluating and validating new analytical methodologies is proven and should be utilized.

KEYWORDS: Routine Analytical Services (RAS), Special Analytical Services (SAS), Contract Laboratory Program (CLP).

INTRODUCTION

The Environmental Protection Agency (EPA) established the Contract Laboratory Program (CLP) in the summer of 1979 to provide state-of-the-art chemical analysis on a high-volume basis, in anticipation of the passage of the Comprehensive Environmental Response, Compensation and Liability Act (SUPERFUND) which was enacted in 1980.

Stanley P. Kovell is the Chief of the Analytical Support Branch of the U.S. Environmental Protection Agency's Office of Emergency and Remedial Response, 401 M Street, S.W. (WH-548A), Washington, DC 20460.

The CLP chose to use private sector laboratories to support the program because the Agency's "in-house" laboratory system could not accommodate the large increase in work associated with the analysis of samples from 50 pre-Superfund "uncontrolled hazardous waste sites" within a six-month period. Through the passage of Superfund, the number of samples increased dramatically. Superfund is the driving force behind the ongoing development of the CLP and is responsible for the program's size and structure.

The purpose of this paper is to describe the organization and infrastructure of the CLP in the context in which it is used in the Superfund methods validation process.

CLP INFRASTRUCTURE

The following section is a description of the CLP infrastructure. Its purpose is to give the reader an understanding of the CLP framework and how it provides an efficient system for Superfund method validation and evaluation.

The CLP developed an infrastructure to efficiently analyze a large volume of samples for use in EPA enforcement cases. Although primarily designed to provide required analytical services, the resulting infrastructure turned out to be very useful for method testing and validation. The CLP consists of numerous Agency groups and support contractors widely distributed throughout the country. The CLP function is performed through the cooperative and interdependent efforts of the Analytical Support Branch of the Office of Emergency and Remedial Response (OERR), the Sample Management Office, the National Enforcement Investigations Center, the Environmental Monitoring Systems Laboratory, the EPA Regions, the EPA Contracts Office, and the contract laboratories. Each of these groups participates in some aspect of method development, review, validation and modification. Figure 1, Interrelationship of CLP Principals, graphically illustrates the interaction of these groups in the CLP operation.

The National Program Office (NPO) routinely directs and coordinates the data review and validation process, which is used to continuously evaluate method performance. The Sample Management Office (SMO) coordinates closely with the Regions, samplers, contract laboratories and CLP management, reports any problems with analytical methods, and participates in the methods modification process. SMO also coordinates the use of new analytical methods for non-routine types of analyses through the Special Analytical Services (SAS) program.

The National Enforcement Investigations Center (NEIC) is the enforcement arm of the CLP. NEIC ensures that the CLP protocols meet enforcement standards so that data generated by the laboratories is legally defensible and can be used as evidence in Agency enforcement actions. NEIC monitors the CLP to ensure that the laboratories maintain sample chain-of-custody throughout the analytical process. NEIC requires each laboratory to rigorously follow documentation standard operating procedures, to ensure that each sample can be tracked through every step in the analytical process. The paper "Document Control and Chain-of-Custody Considerations for the National Contract Laboratory Program," by R.H. Laidlaw, contains detailed information on the role of NEIC in the CLP (6).

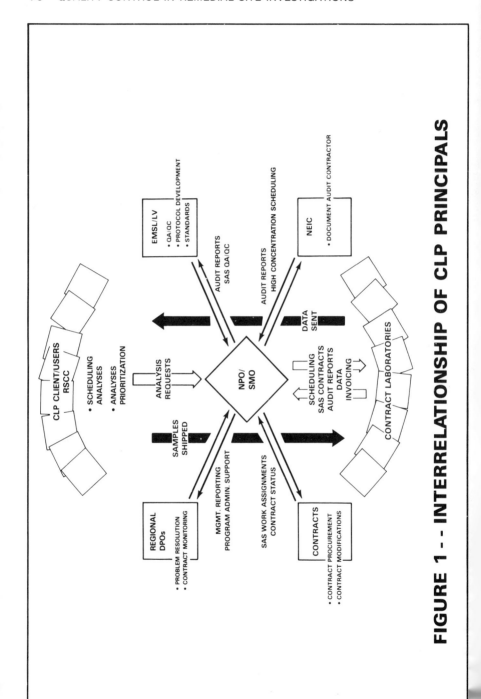

FIGURE 1 - - INTERRELATIONSHIP OF CLP PRINCIPALS

The Environmental Monitoring Systems Laboratory (EMSL/LV) functions as the quality assurance arm of the CLP and plays the key role in the methods validation process. Specifically, EMSL/LV operates the CLP Quality Assurance data base and performs program and laboratory trend analyses used in developing and updating contract Quality Control criteria. In addition, EMSL/LV has developed an extensive system to evaluate laboratory and method performance, using the large CLP data base compiled over the past five years. The results generated are used in method evaluation and modification. Several papers published in STP 925 contain additional information on the role of EMSL/LV in the Superfund methods validation process (1)(2)(3)(4)(5).

The EPA Regions also play an important role in the Superfund evaluation and validation process. Regional data reviewers provide peer review of the analytical methods and the quality of data being generated by providing input on method performance based on data review, and by participating in the technical caucus process. In addition, Deputy Project Officers (DPOs) in each Region work closely with the NPO Project Officer, monitoring laboratory performance and responding to identified problems in laboratory operations. As a result of their close interaction with the laboratories they monitor, the DPOs are instrumental in identifying problems with Superfund analytical methods and play an important role in the methods modification process.

The contract laboratories are another vital component of the validation process. The CLP selects its analytical contractors from the nationwide community of chemical analysis laboratories. CLP laboratories must meet stringent requirements and standards for equipment, personnel, laboratory practices, analytical operations, and quality control operations. The CLP competitively awards firm, fixed-price contracts to the lowest responsive, responsible bidders through the government's Invitation for Bid (IFB) process. Low-priced bidders must successfully analyze performance samples and pass a pre-award laboratory audit before a contract is awarded. EPA closely monitors the laboratories after contract award to assure compliance with the terms and conditions of the contract.

The success of the CLP is dependent on the day-to-day operation of the infrastructure outlined above. The ability of this program to consistently analyze a large volume of samples, while simultaneously validating the methods used in their analyses, is an important result.

PROGRAM INITIATION

In the beginning, the CLP management had no knowledge of who would be using the analytical data produced and what data quality would be required by its clients. In addition, there were no validated analytical methods available for CLP applications, and the precise analytical applications were undefined. It was clear, however, that much of the data generated by this program would certainly be used in litigation where the government would bear a heavy burden of proof. Providing legally-defensible analytical results for use in Agency enforcement actions required the CLP to develop a program with a high level of quality assurance, documentation, and inter- and intra-laboratory consistency.

In order to generate data of high quality and consistency which would be admissible in enforcement cases, the CLP selected the formal advertising process to procure qualified laboratories. The CLP chose this contracting mode because it requires a high degree of specificity in the statement of work, which promotes program consistency and facilitates contract enforcement. The CLP also requires the laboratories to document and provide all data generated in the laboratory as contract deliverables so that any CLP client is able to unilaterally and independently determine the quality of the data without subsequent involvement of the laboratory personnel who actually generated the data. As a result of the structured, disciplined methods specified in the contracts, data generated by CLP laboratories is of known quality and is unequivocally documented.

The CLP first used the 600 series methods, published in the Federal Register (7). Although these methods were not developed for CLP applications, they were the only methods available at the time. The methods were adapted, refined and standardized, and new contracts were awarded periodically, thus implementing state-of-the-art methodologies quickly and efficiently. At the same time data was being generated for program purposes, the CLP was, and is continuing to evaluate methods performance and to modify methods on the basis of "real-world" operating data.

The CLP initiated standardized and specialized analytical services to support a variety of Superfund sampling activities, from those associated with the smallest preliminary site investigations to those of large-scale, complex remedial, monitoring and enforcement actions. The CLP currently operates three separate Routine Analytical Services (RAS) programs: Organic, Inorganic and Dioxin. Standardized organic, inorganic and dioxin analyses are performed by the contract laboratories under the RAS program, using specified Superfund analytical methods. General descriptions of these methods are provided separately in this publication (8)(9)(10)(11).

The RAS program has expanded since the initiation of the CLP in 1979. The number of samples analyzed under the RAS program has grown dramatically over the past five years, as illustrated by Figure 2. In addition, the number of RAS laboratories has increased from 13 in 1980 to 73 in 1985 (see Figure 3).

In 1983, the CLP developed a new service to complement the analyses being performed under RAS. In addition to the standardized analyses and method validation being provided under the RAS program, the Special Analytical Services (SAS) program provides specialized analyses, different from or beyond the scope of the RAS program, but consistent with CLP objectives. SAS services are used whenever client needs cannot be met under the terms and conditions of the RAS contracts. SAS requirements are subcontracted on a case-by-case basis to laboratories performing satisfactorily under the RAS program. Since its inception in 1981, the CLP has used over 200 different methods on a wide range of matrices to meet a variety of non-standard client needs.

The number of samples analyzed through the SAS program has fluctuated greatly the past couple of years and have been increasing recently. Figure 4 illustrates the growth of the Special Analytical

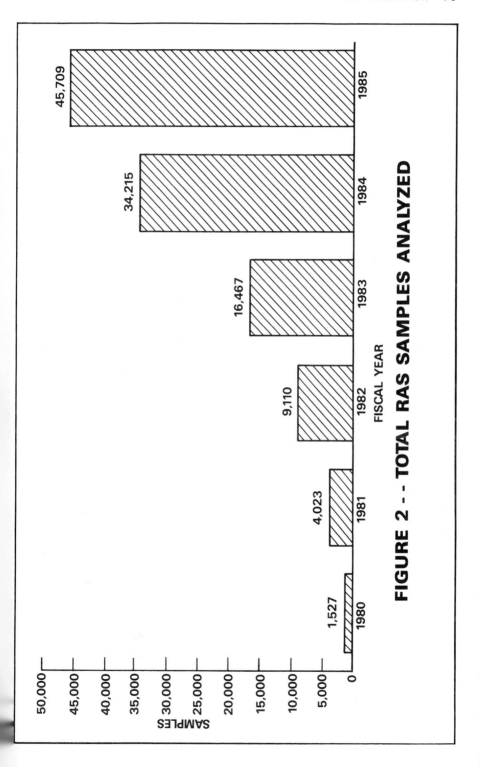

FIGURE 2 - - TOTAL RAS SAMPLES ANALYZED

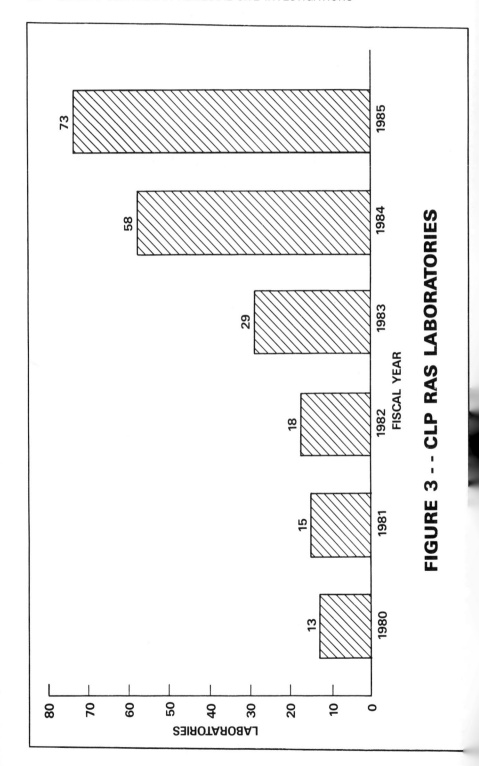

FIGURE 3 - - CLP RAS LABORATORIES

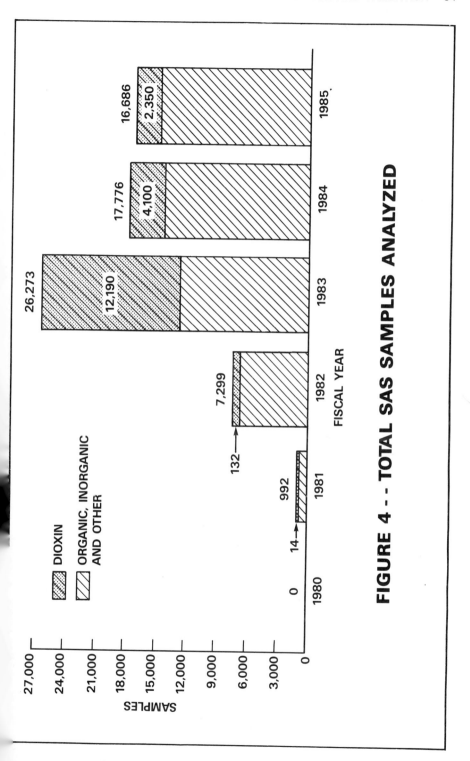

FIGURE 4 - - TOTAL SAS SAMPLES ANALYZED

Services program over the past several years. The requirement to simultaneously provide increased amounts of analytical data involving many new matrices and special methods, created a tremendous methods validation opportunity because it produced an enormous amount of operational data which is vital to the methods evaluation and validation process.

METHOD TESTING

The laboratories currently under contract with the CLP are a large and tested segment of the best environmental analysis, production laboratories in the country. These laboratories are experienced in strictly adhering to the analytical methods specified in their contracts. The laboratories' operational discipline is a key factor in achieving the inter- and intra-laboratory consistency, which is so critical in validating methods in the "real-world" mode.

The CLP uses both RAS and SAS programs for methods validation and evaluation. The RAS program is used to validate a single method being used by a large number of laboratories over a long period of time. The RAS program generates a large data base which EPA uses to evaluate both the laboratories and the methods used to generate the data. The SAS program uses a smaller number of laboratories and the methods testing and evaluation process is usually limited to a single test over a specified time period. As new SAS analytical methods are tested and evaluated, the CLP incorporates them into the RAS program, where the methods are continuously tested and validated by a large number of laboratories.

As a result of the laboratory resources available and the established program structure, the CLP has a viable system to test any new methods with consistency and discipline by the very laboratory community that will be using the methods. In addition, the methods are tested under "real-time" conditions as the samples are being analyzed and data is being generated. The CLP has accepted the operating principle that the inherent variability among the laboratory community, as well as the variability in sample matrices is an integral part of method performance. While this may be a departure from traditional thinking, the enormous variation in matrix from sample to sample renders the "consistent matrix" assumption implicit in traditional methods validation practices, irrelevant for CLP applications. Once a method has been tested through SAS contracting and accepted for general use, the matrix variations from real samples come into play and are accepted as an unavoidable variable in the continuous evaluation of method performance under "real-world" conditions.

METHOD EVALUATION

The technical caucus process is the primary mode used to evaluate analytical methodologies. Technical caucuses are held on a regular basis (usually quarterly) and involve participation by experienced analytical chemists from EPA laboratories, contract laboratories, program support contractors, SMO, EMSL/LV, NEIC, OERR and others as appropriate. The top technical experts from the CLP are assembled at the caucuses to review and refine specific Superfund methodologies. These caucuses have been successful in identifying potential methods performance problems and in recommending corrective changes in methodology. This ongoing evaluation of analytical methods is designed to correct method problems and to incorporate new technology.

The quality assurance/program monitoring role being performed by the Environmental Monitoring Systems Laboratory in Las Vegas is another very important aspect of the method evaluation process. This laboratory has developed the extensive systems that are currently being used to continuously evaluate CLP laboratory and analytical methods performance. EMSL/LV has developed performance data bases for over 80 laboratories for five years using well documented methods applied to the analysis of over 100,000 samples in various matrices. Over 300 methods and matrix variations have been utilized under the SAS program. The experience gained from these activities is directly applicable to evaluating any new methodology used by the laboratories in the CLP system. Detailed information on the method evaluation process is provided separately in this publication (1)(2)(3)(4)(5).

CONCLUSION

The potential for utilizing the CLP for evaluating and validating new analytical methodologies is proven. This testing has clearly reduced the time between research and practical applications of new analytical methods. In addition, the CLP is currently applying its process to validating Resource Conservation and Recovery Act (RCRA) methods. The sharing of results from method evaluations with others in the scientific community will further the development and refinement of new analytical procedures.

In conclusion, the CLP is engaged in an empirical process of methods validation and evaluation of analytical methods currently in use. The CLP offers a unique opportunity to test new methodologies under "real-world" conditions by the laboratory community that will ultimately be using the methods. This new approach to method validation and evaluation should be utilized.

ACKNOWLEDGEMENT

The author acknowledges with appreciation the assistance of Steven Manzo, Sample Management Office, in the preparation of this paper.

REFERENCES

(1)   Flotard, R.D., Homsher, M.T., Wolff, J.S., and Moore, J.M., "Volatile Organic Analytical Methods — Performance and Quality Control Considerations," Quality Control in Remedial Site Investigation: Hazardous and Industrial Solid Waste Testing, Fifth Volume, ASTM STP 925, C.L. Perket, Ed., American Society for Testing and Materials, 1986.

(2) Marsden, P.J., Pearson, J.G., and Bottrell, D.W., "Pesticide Analytical Methods — General Description and Quality Control Considerations," Quality Control in Remedial Site Investigation: Hazardous and Industrial Solid Waste Testing, Fifth Volume, ASTM STP 925, C.L. Perket, Ed., American Society for Testing and Materials, 1986.

(3) Wolff, J.S., Homsher, M.T., Flotard, R.D., and Pearson, J.G., "Semivolatile Organic Analytical Methods Performance and Quality Control Considerations," Quality Control in Remedial Site Investigation: Hazardous and Industrial Solid Waste Testing, Fifth Volume, ASTM STP 925, C.L. Perket, Ed., American Society for Testing and Materials, 1986.

(4) Garner, F.C., Homsher, M.T., and Pearson, J.G., "Performance of USEPA Method for Analysis of 2,3,7,8-Tetrachloro-dibenzo-p-dioxin in Soils and Sediments by Contractor Laboratories," Quality Control in Remedial Site Investigation: Hazardous and Industrial Solid Waste Testing, Fifth Volume, ASTM STP 925, C.L. Perket, Ed., American Society for Testing and Materials, 1986.

(5) Aleckson, K.A., Fowler, J.W., and Lee, Y.J., "Inorganic Analytical Methods Performance and Quality Control Considerations," Quality Control in Remedial Site Investigation: Hazardous and Industrial Solid Waste Testing, Fifth Volume, ASTM STP 925, C.L. Perket, Ed., American Society for Testing and Materials, 1986.

(6) Laidlaw, R.H., "Document Control and Chain-of-Custody Considerations for the National Contract Laboratory Program, Quality Control in Remedial Site Investigation: Hazardous and Industrial Solid Waste Testing, Fifth Volume, ASTM STP 925, C.L. Perket, Ed., American Society for Testing and Materials, 1986.

(7) Federal Register, Monday, December 3, 1979, Friday, October 26, 1984.

(8) Kleopfer, R.D., "Dioxin Analytical Methods — General Description and Quality Control Considerations," Quality Control in Remedial Site Investigation: Hazardous and Industrial Solid Waste Testing, Fifth Volume, ASTM STP 925, C.L. Perket, Ed., American Society for Testing and Materials, 1986.

(9) Fisk, J.F., "Volatile Organic Analytical Methods — General Description and Quality Control Considerations," Quality Control in Remedial Site Investigation: Hazardous and Industrial Solid Waste Testing, Fifth Volume, ASTM STP 925, C.L. Perket, Ed., American Society for Testing and Materials, 1986.

(10) Fisk, J.F., "Semi-Volatile Organic Analytical Methods — General Description and Quality Control Considerations," Quality Control in Remedial Site Investigation: Hazardous and Industrial Solid Waste Testing, Fifth Volume, ASTM STP 925, C.L. Perket, Ed., American Society for Testing and Materials, 1986.

(11) White, D.K., "Inorganic Analytical Methods — General Description and Quality Control Considerations," Quality Control in Remedial Site Investigation: Hazardous and Industrial Solid Waste Testing, Fifth Volume, ASTM STP 925, C.L. Perket, Ed., American Society for Testing and Materials, 1986.

John M. Moore and J. Gareth Pearson

QUALITY ASSURANCE SUPPORT FOR THE SUPERFUND CONTRACT LABORATORY PROGRAM

REFERENCE: Moore, J. M. and Pearson, J. G. "Quality Assurance Support for the Superfund Contract Laboratory Program," Quality Control in Remedial Site Investigation: Hazardous and Industrial Solid Waste Testing, Fifth Volume, ASTM STP 925, C. L. Perket, Ed., American Society for Testing and Materials, Philadelphia, 1986.

ABSTRACT: The Environmental Protection Agency's Environmental Monitoring Systems Laboratory-Las Vegas (EMSL-LV) functions as the quality assurance (QA) arm of the Contract Laboratory Program (CLP). The EMSL-LV provides standards and quality assurance materials, maintains a quality assurance data base, and conducts data audits, performance evaluation studies, and on-site laboratory evaluations. In addition, the EMSL-LV is very actively involved in evaluating laboratory and method performance. Due to its unique position of receiving all of the data from all of the CLP laboratories, combined with its charge of quality assurance oversight, EMSL-LV is most able to assess intra- and inter-laboratory performance as well as programwide method performance. This paper describes the EMSL-LV QA program for the CLP as of October 1985.

KEYWORDS: quality assurance, Superfund, Contract Laboratory Program, standards, data base, performance evaluation, data audits, laboratory evaluations.

INTRODUCTION

Background and Scope

Since its inception in fiscal year (FY) 1980, the number of laboratories in the Contract Laboratory Program (CLP) has been increasing continuously in response to the need for additional analytical data.

Mr. Moore is the Manager for Data Audits/On-site Evaluation Program, and Mr. Pearson is the Branch Chief of the Toxics and Hazardous Waste Operations Branch of the Quality Assurance Division at the Environmental Monitoring Systems Laboratory, P.O. Box 15027, Las Vegas, NV 89114.

Concurrently, there has been a growing recognition of the need to (1) improve the ability to determine data quality and (2) improve the means of defining data quality. Procedures used to provide QA support have been modified, and new procedures have been developed to accommodate current and projected growth. These improvements will continue because of the dynamic state of the CLP and its QA needs. The scope of this paper covers the QA responsibilities and activities of EMSL-LV that support the CLP in general. The paper specifically focuses on those responsibilities and activities of EMSL-LV that concern the contract laboratories and their analytical output. This paper does not address non-analytical aspects of QA, such as sampling and preservation, shipping, custody, document control or other matters outside the responsibility of EMSL-LV.

The objectives of the EMSL-LV QA effort are to:

1. Define the quality of data produced.
2. Improve the quality of data.
3. Monitor technical compliance with contractual QA requirements.
4. Evaluate laboratory performance.
5. Evaluate method performance.

The QA effort consists of several major elements, which are discussed in detail in ensuing sections of this paper. They are:

1. Analytical Standards and Reference Materials: Design, preparation, characterization, inventory maintenance and distribution.
2. Quality Assurance Data Base: Development and updating of QA and Quality Control (QC) criteria, data storage, trend evaluation and QA report preparation.
3. Laboratory Evaluation and Performance Monitoring: On-site evaluations, data audits and performance evaluation (P.E.) studies.

Laboratory evaluation and performance monitoring consist of two major activities:

1. Preaward assessment of a laboratory's (a) ability to meet all contract requirements and (b) sample volume throughput capacity.
2. Postaward performance monitoring, data acquisition, problem identification and correction of deficiencies.

Overview

The objective of the CLP is to meet the Superfund needs of the EPA Regions, States and other agencies for analytical information of high quality through a decentralized, standardized system. The major organizational elements of this national program are the National Program Office (NPO), Sample Management Office (SMO), EMSL-LV, the National Enforcement Investigations Center (NEIC), contract laboratories and the clients (users) of the Program.

The infrastructure of the CLP is described in detail in "Evaluating and Validating Analytical Methods Using the Environmental Protection Agency's Contract Laboratory Program," by S. P. Kovell [1]. Very

briefly, the NPO has overall management responsibility of the CLP; the
SMO provides administrative and sample control functions; EMSL-LV provides quality assurance; NEIC is responsible for documentation and
chain-of-custody aspects; the contract laboratories deliver analytical
chemistry analysis to the CLP clients, primarily the 10 EPA Regional
Offices.

Contract Laboratory Program: The CLP now involves approximately
75 laboratories in all 10 EPA Regions.   These laboratories perform
inorganic, organic and dioxin analytical services as required by the
EPA Regions, as well as other users of the Program.   The methods are
explicitly defined in the protocols for the organic, [3] inorganic,
[4] and dixoin [5] contracts.

The approximate number of laboratories engaged in each category
of analysis is:

Analysis Category	Number of Labs
Organics	44
Inorganics	18
Dioxin	13

Through the Invitation for Bid (IFB) process, contracts are awarded
to commercial laboratories submitting the lowest, responsive, responsible bids.   A unique system for selection of analytical laboratories
was developed which includes very rigorous quality control requirements.

The system includes the following activities:

1.   Evaluation of each laboratory's understanding of its contractual requirements.
2.   Assessment of each laboratory's capability to satisfy those
contractual specifications.
3.   Assessment of each laboratory's capacity to produce an adequate flow of data, suitable for enforcement case preparation and
litigation.
4.   Continuous monitoring of performance and assurance of compliance with the contractual requirements.

The laboratories currently in the CLP have undergone the award
process outlined in Table 1.   This process provides a reasonable
measure of confidence that the standards of performance specified in
each contract will be achieved.

ANALYTICAL STANDARDS AND QUALITY ASSURANCE MATERIALS

Overview

Standards and reference materials of known purity are made available to all CLP participants to be used for QC operations and to

provide a basis for uniformity between laboratories with regard to instrument calibration and analyte identification by establishing traceability to a common source of analytical standards.

TABLE 1--CLP Contract award process.

1. EPA Program Office - Based on user requirements, establishes need to:
   a. Provide additional analytical capacity.
   b. Develop new or revised analytical requirements/protocols.
2. EPA Contract Office - Announces solicitation.
3. EMSL-LV:
   a. Sends P.E. a performance evaluation sample set to laboratories seriously interested in the contract.
4. Contract Laboratories:
   a. Analyze P.E. sample set.
   b. Submit P.E. results to EMSL-LV.
   c. Submit bids to Contract Office.
5. EMSL-LV:
   a. Evaluates and scores P.E. sample set results.
   b. Provides P.E. sample results to Contract/Program Offices.
6. Contract/Program Offices:
   a. Receive and open bids.
   b. Identify lowest-bidding laboratories.
7. EMSL-LV/Program Office:
   a. Recommend which laboratories are ineligible for further consideration, based on P.E. score.
   b. Schedule preaward on-site laboratory evaluations.
   c. Assist Contract Officer in conducting on-site evaluations, assesses laboratory sample capacity and ability to meet the terms and conditions of the contract.
8. EMSL-LV - Makes recommendations to Program/Contract Offices for award (or non-award) and estimates laboratory's sample capacity.
9. Program/Contract Offices - Determine award and capacity obligation of laboratory.
10. Contract Office - Awards contract.

The CLP interacts with an EPA-wide program that has established the Quality Assurance Reference Materials Project (QARMP). This project is jointly managed by EMSL-LV and Environmental Monitoring Support Laboratory-Cincinnati (EMSL-CI), and is supported by a contractor, currently Northrop Services, Inc. (NSI). The functions, responsibilities and mechanisms of this effort are detailed in a Memorandum of Understanding (MOU) between the two EMSL's. This MOU establishes, for management purposes, two interdependent entities [2]:

1. EPA Repository for Toxic and Hazardous Materials (EMSL-CI)
2. Quality Assurance Materials Bank (QAMB) (EMSL-LV)

The interrelationships of the Repository and the QAMB are outlined in Fig. 1.

Repository: This activity is directed by EMSL-CI which is responsible for verified neat compounds and calibration standards. This includes:

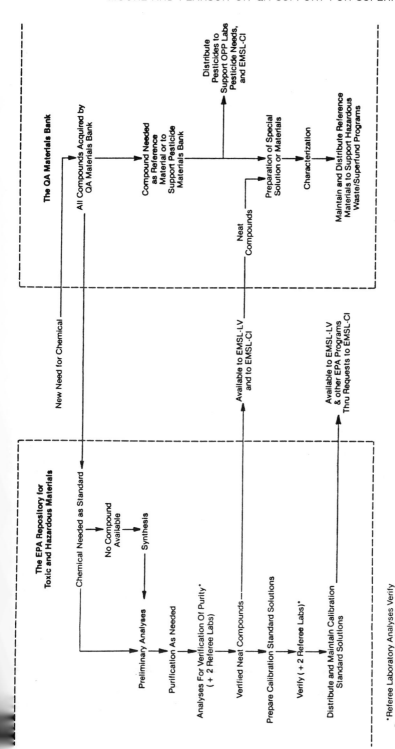

FIG. 1--Interaction of EPA repository and QA materials bank.

a. Acquisition of reference materials from the QAMB and analysis to determine their purity.

b. Purification of the reference materials if necessary.

c. Synthesis of compounds that are unavailable or too expensive from commercial sources.

d. Analysis of the purified neat compounds.

e. Preparation of purified compounds in organic solvents for use as calibration standards and spiking solutions. Also, analysis of those substances to determine purity and/or concentration.

f. Maintenance and distribution of an inventory of verified neat compounds and calibration standards.

Pure, neat compounds, calibration standards and spiking solutions are also analyzed by referee laboratories prior to their use.

Quality Assurance Materials Bank: This activity is directed by EMSL-LV which is responsible for reference materials and the following activities:

a. Acquisition of materials from commercial sources for Repository use.

b. Acquisition of materials from the Repository for use in EMSL-LV programs.

c. Characterization of materials (chemicals or mixtures) for EMSL-LV programs.

d. Preparation of sets of reference materials in ampuls for P.E. studies to support the CLP.

e. Maintenance of an inventory and distribution system for these materials.

Quality Assurance Materials--Organics

The protocol for the analysis of organic pollutants specifies the compounds required to conduct all calibration, analytical and QA/QC operations. These materials include compounds on the Superfund Hazardous Substances List (HSL), surrogates, internal standards and matrix spike materials. (See Table 2 for a breakdown of the number by fraction.)

TABLE 2--Number of hazardous substances, surrogate spikes, internal standards, and matrix spikes associated with CLP organic protocol.

Fraction	HSL	Surrogates	Internal Standards	Matrix Spikes
VOA's	35	3	3	5
Semi-VOA's	65	6	6	12
Pesticides	26	1	N/A	6

In general, the need for these materials is established by the requirements of the protocol. Requests for needed organic QA materials and suggestions for new materials come to EMSL-LV from all organizations in the CLP community. Priorities are established by EMSL-LV.

Specifications for new materials, such as multicomponent standards previously unavailable, are made by EMSL-LV.

At present, all compounds are available in solutions of known purity and concentration. Surrogates, internal standards and matrix spike materials are also available. Some materials are available in the form of multicomponent mixtures.

Examples of single- and multi-commponent standards inventory maintained to support the CLP through the QAMB are shown in Tables 3 and 4, respectively.

TABLE 3--Sample of single-component standards inventory.

Std. Code	Primary Name	Solvent	Conc.	Purity	Ampuls	Date
16604	Nitrobenzene-d_5	Methanol	5,000	99+	50	02/83
17901	2,4,5-Trichlorophenol	CH_2Cl_2	1,000	99+	400	12/82
20603	2,3,7,8-TCDD	Isooctane	7.87	99+	400	01/83
23003	Phenanthrene-d_{10}	CH_2Cl_2	1,000	99+	300	11/82
23202	Fluorobenzene	Methanol	5,000	99+	100	08/82
28402	Acetone	Tetraglyme	5,000	99+	60	12/82
30502	4-Chloroaniline	Benzene	1,000	99+	400	12/82
33601	2-Hexanone	Methanol	5,000	99+	120	12/82

TABLE 4--Multicomponent standards example - Mixed Acids.

Standard Code: C-016   Solvent: Methanol

Repository Number	Compound	Purity	Concentration $\mu g/mL$
E-000022	2-Chlorophenol	99.1	1,000
E-000029	2,4-Dichlorophenol	99.5	1,000
E-000224	2,4-Dimethylphenol	97.6[a]	1,000
E-000058	4,6-Dinitro-o-cresol	99.3	1,000
E-000056	4-Nitrophenol	99.3	1,000
E-000019	2,4,6-Trichlorophenol	99.3[a]	1,000

[a] Concentration corrected for purity.

Quality Assurance Materials--Inorganics

At this time no certified reference materials or analytical standards are available through the CLP, since no materials have been obtained or prepared specifically for this program.

Currently the EMSL-LV, through a contract with the University of Nevada-Las Vegas, provides an ICP Interference Check Solution. In addition, the following types of materials are being incorporated into CLP: (1) initial calibration verification solutions, and (2) liquid and solid laboratory control samples.

For analytical calibration materials, the common practice appears to be the use of commercially prepared stock solutions (i.e., AA standards) for the elements by preparing working dilutions from individual element concentrates (typically 1,000 ppm). For inductively coupled argon plasma (ICAP) calibration, a series of compatible multielement mixtures are prepared at varying elemental concentrations. For AA furnaces and cold vapor elements, working calibration dilutions are prepared for each element. For cyanide, working calibration standards are prepared from stock reagent grade chemicals. Verification of calibration standards is frequently accomplished by using water/wastewater multielement QC check samples obtained from EMSL-CI as an initial calibration verification check sample. As indicated above, the EMSL-LV will soon be providing initial calibration verification check samples that are protocol specific.

Quality Assurance Materials--Dioxin

At the present time, all materials for dioxin analyses are available from the EMSL-LV, including calibration standard solutions, internal standard and surrogate compounds, column performance check mixtures and fortification solutions. These materials are provided through a contract with Northrop Services, Inc.

QUALITY ASSURANCE DATA BASE

Overview

The QA Data Base (defined below) is the focal point for QA/QC data storage. A number of computer programs provide access to the cumulative data files, and permit identification of trends, evaluation of QC criteria and the updating and development of performance-based QC criteria. The data are obtained from laboratory QA/QC reports, P.E. and Preaward Evaluation reports and onsite evaluation summaries.

The Data Base (used here as a generic, collective term) actually consists of three separate data bases -- organics, inorganics and dioxin. Each has its own analysis and reporting system. Until recently, the Data Base was maintained on the EMSL-LV PDP 11/70 computer. Due to several factors, notably performance constraints and the Agency-wide need for access to CLP information, the Data Base was transferred to the IBM 3081 computer at EPA's National Computer Center (NCC) at Research Triangle Park, NC.

Data packages are inventoried upon receipt, and QC data are organized by EPA Region, contract laboratory and case and sample numbers. Data entry is accomplished manually via input screens from standard forms. There is a master file for quality control parameters for all three programs. In addition, sample result data for the dioxin protocol are in the Data Base.

From the Data Base, internal laboratory control and relative laboratory performance can be evaluated. Performance trends and defects can then be monitored within a given laboratory or between laboratories. Performance-based QC criteria (e.g., acceptance windows for surrogate spike recoveries) can be evaluated and updated as needed.

Data Base QA Reports

A number of QA reports from the Data Base are produced to support the CLP. These reports provide information on QA/QC criteria and performance, organized by analytical parameter, laboratory, sample, matrix and Region. These reports are provided to a variety of users and the CLP laboratories themselves.

Data Base QA Criteria

The Data Base has provided the information needed to establish performance-based criteria in updated analytical protocols, where advisory criteria had been previously. The CLP program produces an enormous amount of analytical data from a large set of independent, commercial laboratories. This vast empirical data set is carefully analyzed, with the results augmenting theoretical and research-based performance criteria. The result is a continuously monitored set of quality control and performance criteria specifications of what is routinely achievable and expected of state-of-the-art analytical chemistry laboratories in mass production analysis of acutely sensitive environmental samples. This, in turn, assists the Agency in meeting its objectives of obtaining data of known and documented quality.

Organics: The criteria for the QA/QC actions are specified in the protocol, Exhibits B and E of the contract [3]. Special forms are provided for recording the required data. These forms are submitted with each data package, along with the other required sample data. These forms provide for direct Data Base entry of:

1. Surrogate percent recovery, water and soil.
2. Matrix spike and duplicate recovery, water and soil.
3. Method blanks.
4. GC/MS tuning and mass calibration, BFB and DFTPP.
5. Initial calibration data.
6. Continuing calibration check data.

Inorganics: The inorganics protocols detail the QA/QC criteria in Exhibits B and E of the contract [4]. Special forms are included for the recording of QC data required with each data package. These data are subsequently entered into the Data Base. The QC data forms provide for the entry of:

1. Initial and continuing calibration.
2. Method blanks.
3. ICP interference check sample.
4. Matrix/digestion spike recovery.

5. Duplicates.
6. Instrument detection limits/laboratory control sample.
7. Standard addition results.

Dioxin: The Dioxin results and QA/QC criteria deliverables are
specified in Appendix B of the contract [5]. These forms permit Data
Base entry of the following:

Form No.	Data
B-1	TCDD sample data, 320/322 ratios, 332/334 ratios, isotopically labeled surrogate recovery, relative ion abundance
B-2	Initial calibration summary, native surrogate, TCDD isomer resolution, duplicate data
B-3	Continuing calibration summary (see B-2, above)
B-4	Partial scan confirmation response ratios, percent relative ion abundance

Reports: Reports from the Data Base include control charts,
exception reports, trend analyses, summary reports, and statistical
analyses. In the case of dioxin, reports are also produced on the
analytical results. These reports are used to examine within and
between laboratory performance, as well as to evaluate method
performance.

LABORATORY PERFORMANCE EVALUATION AND MONITORING

Overview

     The heart of the QA support effort is the evaluation and moni-
toring of laboratory performance. The evaluation and monitoring of
performance has two major parts:

1. Preaward: To assess a laboratory's analytical capacity
    and its qualifications and capability to meet Program
    requirements.
2. Postaward: To monitor, evaluate, and improve the laboratory's
    technical performance and contract compliance, and to identify
    and correct problems.

     These activities are implemented by the interaction of three major
functional areas:

1. Data Audits.
2. Performance Evaluation Studies.
3. On-site Laboratory Evaluations.

     Described here are the procedures and tools developed and utilized
by EMSL-LV to evaluate and monitor laboratory performance. All general

and specific criteria in the contract protocols are addressed by those procedures [6-9].

By addressing these criteria through the functions of Data Audits, P.E. Studies and On-site Evaluations, it is possible not only to detect the existence of a problem, but to characterize it and trace its source. Such a problem may or may not be laboratory specific. The problem may be identified as a weakness in the analytical method, an ambiguity in the protocol language or a problem with a particular analyte.

The evaluation and monitoring process has adapted to the changing needs of the CLP due to the growth in the number of laboratories, the concurrent increase in the quantity of data and the necessity of developing more objective and efficient means of measuring data quality.

Data Audits

Overview: The Data Audit function is managed by the EMSL-LV. Data Audit activities include:

1. Technical audits of contract laboratory data with respect to contract requirements.
2. Monitoring and consulting to update and improve procedures, systems, and contracts.
3. Updating the QA Data Base.
4. Developing documentation--checklists, control charts and scoring systems.
5. Performing special audits (preparation for litigation).
6. Participating in on-site evaluations.
7. Providing material for technical presentation and/or publications.
8. Providing statistical support to all aspects of the CLP.
9. Providing expert witness testimony.
10. Performing methods evaluation.

The technical audits are the linchpin of all the above activities. They provide a detailed working knowledge of what is actually being accomplished in the CLP laboratories. This knowledge base becomes the reference from which all other activities are measured and establishes the credibility of all laboratory and method performance activities.

The technical audits are extremely thorough and consequently are conducted on only a small, but statistically valid, fraction of all the cases. (A case is a collection of environmental samples from one site over a finite period of time.)

Military Standard 105D (MIL STD 105D), which is a general inspection and statistical sampling method, has been adapted to the Data Audit process. This method establishes, on a laboratory-by-laboratory basis, valid statistical levels of data inspection, based on the quantity of data reported and relative frequency of observed defects. It also provides a means for random selection of data to be audited, thus resulting in a less-selective, unbiased characterization of a laboratory's output. By utilizing the scoring system applied, scores are weighted based on the severity of observed defects, and the appropriate level of data inspection can be projected. This adaption of MIL STD 105D is derived from the original document [10] as described in an interim report [11].

Data audits--detailed and sample-specific: Data audits at the appropriate level of inspection can flag details that indicate the need for corrective action, but which would not otherwise be identified by the routine computer entry of QA/QC data.

Data Audit forms have been developed and are in use for the routine review and audit of laboratory data packages on a case-specific basis for organics, inorganics and dioxin contracts. These forms, designated as "plates," provide for the assimilation of detailed information which allows errors to be detected and operational and other deficiencies to be identified.

The Data Audit Plates now in use incorporate the application of MIL STD 105D. They provide for the classification of observed defects as critical, major or minor. A scoring system is established based on the nature and number of defects identified in a given audit. The types of defects are defined as follows:

1. Critical--affects validity of data of entire case, such as missing major forms, blank data, etc.
2. Major--affects a sample for all parameters or a given parameter throughout the case.
3. Minor--affects a given parameter for a sample or a readily resolvable problem relative to general technical content.
4. Method--affects a given compound or parameter from a sample and is not resolvable between a laboratory or method problem.

Between-laboratory data audits: Between-laboratory audits provide an informational link between Data Audit and P.E. Studies. Printouts are available of preaward or routine P.E. results, by analyte and laboratory, showing mean values, true values, standard deviations and confidence intervals. This enables one to compare analytical performance of a given laboratory with respect to either all laboratories or to any other single laboratory.

Data Base Summary reports are another example of the use of Data Audits to assess laboratory-to-laboratory performance. In this report, a particular spiking compound or element may be selected, and the mean recovery (by matrix and laboratory) can be compared with the overall results obtained throughout the entire CLP.

A laboratory's poor performance, relative to that of other laboratories, can be assessed by using these reports. The reports provide concrete facts that can make the resolution of problems at on-site evaluations more productive. Also, the data in the reports can also be used to detect a general, CLP-wide problem.

Performance Evaluation Studies

Overview: Performance Evaluation (P.E.) studies assess laboratories' analytical performance on "blind" samples prepared by the
EMSL-LV. There are two types of P.E. studies: preaward and postaward.

The preaward analyses are used to evaluate a laboratory's ability
to correctly identify and quantitate analytes covered in the protocols.
The resulting scores from the analysis of these samples are used as
part of the information required to determine if that laboratory should
be considered further for a contract award. By their very nature, preaward P.E. studies are conducted on an "as-needed" basis.

The postaward, or routine, P.E. studies are conducted quarterly
and are a means of continuously monitoring the quality of laboratory
data with respect to the protocol requirements. There is growing
interest in performing these studies on double blind samples, and this
is currently done on aqueous organic samples and on inorganic aqueous
and soil samples. Dioxin P.E. studies are conducted as single blinds.
There is also interest in developing appropriate solid matrices and
using spiked, real environmental samples to make samples less recognizable. This would permit P.E. studies to more accurately determine how
a laboratory routinely performs.

References are made to single blind and double blind P.E. samples.
These are defined as follows: single blind—a sample that is recognizable as a P.E. sample, but the contents of which are unknown to the
analyst. Double blind—a P.E. sample which is indistinguishable as such
from regular field samples, and the contents are obviously unknown to
the analyst.

Preaward Performance Evaluation Studies

Organics: The preparation of preaward materials is triggered by
the issuance of a new contract solicitation. The EMSL-LV reviews the
protocol to determine what materials are needed to perform an analysis
in accordance with the protocol requirements.

In addition to the single blind samples prepared for analysis,
each laboratory must be provided with a complete package of traceable
calibration standards, surrogates, internal standards, tuning compounds,
etc., required by the protocol. Samples are designed to contain representative HSL compounds at appropriate target concentration levels.
The materials needed are then obtained from the QAMB. Complete sets
of ampulated samples, standards, etc., are packaged and sent, with
appropriate instructions, to the laboratories from the EMSL-LV. The
identical sample set is sent concurrently to several referee laboratories for analysis.

After the data packages are returned to EMSL-LV, the packages are
reviewed for QA/QC compliance and completeness, and the analytical
scoring forms for each laboratory are completed. These forms account

for the detection, identification and quantitation of analytes, QA, tuning, screening and deliverables. Results of the preaward scoring are provided to the Program/Contract Offices along with recommendations for on-site laboratory evaluations.

Inorganics: The preaward procedure for inorganics is very similar to that for organics.

The standards used for the inorganic single blind samples are generally the best available commercial materials, such as high purity metal salts. The solid matrix may be a clean sand or natural soil matrix. Each P.E. sample is prepared in a single batch, ampulated and packed into individual sets. They are distributed in the same manner as the organic P.E. sample sets.

Scoring and any recommendations for on-site laboratory evaluations are similar to the organic preaward procedure.

Dioxin: The preaward sequence for dioxin contract solicitations is similar to that for organics and inorganics. The dioxin P.E. sample set includes a solid matrix blank (clay) and a series of other materials sufficient to evaluate laboratory performance in relation to all analytical aspects of the protocol.

The resulting single blind data are reviewed in accordance with details specified in the contract. Any false positive 2,3,7,8-TCDD reported may be grounds for disqualification of a laboratory for contract award consideration.

Scoring and any recommendations for on-site laboratory evaluations are similar to the organic preaward procedure.

Routine Performance Evaluation Studies

Organics: The samples to be used in postaward P.E. studies are generally defined by EMSL-LV, but the specific design and preparation is executed by the Environmental Monitoring and Support Laboratory-Cincinnati (EMSL-CI). At present, there is no solid matrix post-award organic P.E. sample.

The P.E. samples are prepared, analyzed and shipped by the EMSL-CI They are shipped (1/2-gallon aliquots plus VOA vials) directly to an EPA Regional laboratory. The sample sets contain field water which has been diluted with laboratory pure water, and the mixture is then define as the "field" Blank. They are then submitted by the EPA Region to the laboratories as individual sample cases. When samples are to be submitted as single blinds, they are shipped directly to the laboratories along with instructions prepared by EMSL-LV.

The scoring is accomplished by the EMSL-LV. In evaluating the P.E scores, if the data package from a given laboratory is complete, and the score is high, no full data audit is performed. If the score is low, a full data audit may be performed. A low P.E. score is used as an indicator of the need for an on-site visit. The exceptions listed on the score sheet and resultant conclusions and recommendations are included in a report to each laboratory, informing it of its performance. This letter also goes to the appropriate EPA Project Officer, who implements any recommended action required.

Poor performance on a P.E. sample may result in a show-cause order from the Contract Office, which stops the flow of samples to the laboratory pending the implementation of corrective actions that are identified and discussed at an on-site evaluation.

Inorganics: The sample types and materials and procedures for design, specification and production of the materials are identical to those described for inorganics preaward P.E. samples. Generally, water and soil samples are now sent to the EPA Regions for submission as double blind P.E. samples. The data from these samples is evaluated and reported in the same manner as the organic P.E. data described above.

Dioxin: The procedures used to routinely evaluate dioxin analytical performance differ substantially from those used in organics and inorganics contracts. Dioxin programs do not evaluate performance on a quarterly basis, but provide for on-going, continuous monitoring by requiring the analysis of a P.E. sample with each case of samples analyzed.

The samples themselves are the same type as those used for preaward evaluations. There is an array of P.E. samples available at various concentrations. They are distributed by the EMSL-LV to the EPA Regions in 20-g portions. The Regions submit them with field samples to the laboratories. These P.E. samples also may be spiked with interfering compounds to test cleanup and resolution procedures.

Analysis of the P.E. result permits the assessment of performance and evaluation of trends on a within-laboratory and between-laboratory basis. Full data audits then can be conducted if necessary.

On-Site Laboratory Evaluations

Overview: The on-site laboratory evaluation is the QA support activity which ties together all the elements of laboratory performance monitoring. Preaward on-site evaluations, together with P.E. sample results, are critical factors in the determination of contract awards. Routine on-site evaluations incorporate the findings of P.E. studies and data audits to detect and identify problems, bring the problems to the attention of the cognizant laboratory personnel, help to resolve them and otherwise monitor for compliance with all contractual specifications. If serious problems are encountered, recommendations are made to the Program Office for appropriate corrective action.

Preaward on-site evaluations: The Contract Officer notifies the qualified laboratories that preaward on-site evaluations will occur. The principal objectives of the on-site evaluation are:

1. Verification of the claims made in the laboratory's bid.
    This is to ensure that the laboratory facilities (instrumen-
    tation, bench space, hoods, glove boxes, storage areas,
    document support equipment, etc.) exist and are functioning
    as claimed by the laboratory. Key personnel are identified
    and their qualifications are reviewed.

2. Verification of the laboratory's ability to comply with
    contractual requirements. Determination is made that the
    basic needs for recognizance and accountability are adequate,
    and that procedural systems adaptable to the contract require-
    ments are in place (e.g., internal QA/QC, sample tracking
    chain-of-custody documentation, data review, and SOP's for
    general laboratory operations).

3. Determination of capacity. This determination, of course, is
    related to the number of samples the laboratory bids for and
    the number available for award. The factors described above
    also influence this assessment. In addition, instrument
    redundancy is required for organics and dioxin contracts to
    guarantee that if an instrument needs repair, reporting dates
    can still be met. Other factors, such as storage space, hood
    space and refrigerators, etc., are also considered.

An outline of the sequence of events and the items covered at a
preaward on-site evaluation is shown in Table 5.

TABLE 5--Standard procedure for preaward on-site evaluation.

1. Meeting with Laboratory Manager and Project Manager
2. Verification of Personnel
3. Verification of Instrumentation and General Laboratory Facilities
4. Review of Quality Control Procedures
5. Review of Standard Operating Procedures (SOPs)
6. Review of P.E. Sample Results
7. Identification of Needed Corrective Actions

Routine On-Site Evaluations: As described earlier, the objective
of preaward on-site evaluations is to determine each laboratory's
potential ability to comply with the contractual requirements and to
estimate its sample capacity. Routine on-site evaluations monitor its
progress and adherence to those requirements.

The on-site evaluations are managed by EMSL-LV. The evaluation
team consists of (1) a team leader, who is a senior EMSL-LV staff
member, (2) generally one or two members of the EMSL-LV contract support
staff and/or UNLV-QAL staff serving as technical experts and (3) a
representative from the EPA National Enforcement Investigation Center
(NEIC) to deal with documentation and chain-of-custody matters. Repre-
sentatives of Program Management Contract Office or EPA Regions may
also be present, depending on the prevailing circumstances.

Early in the program, routine on-site evaluations were conducted quarterly. Now that the number of CLP laboratories has more than quadrupled, evaluations are less frequent due to resource limitations. Data audits, P.E. results, and Data Base reports help to identify those laboratories which are moved ahead for an on-site evaluation.

The major topics covered during the postaward, or routine, on-site evaluation are listed in Table 6.

TABLE 6--Standard procedure for postaward on-site evaluation.

1.   Review of Previously Identified Problems
2.   Review of Data Audit Reports
3.   Review of P.E. Study Results
4.   Review of Laboratory QA/QC Procedures and GLPs
5.   Laboratory Tour/Site Audit Check List
6.   Identification of Needed Corrective Action

CONCLUSION

The CLP quality assurance support program is formal, rigorous, and has proven to be successful. And this it must be to achieve a basic EPA premise: that all environmental monitoring data be of known and documented quality.

ACKNOWLEDGEMENTS

Although the research described in this article has been funded by the United States Environmental Protection Agency, it has not been subjected to Agency review and therefore does not necessarily reflect the views of the Agency, and no official endorsement should be inferred. Mention of trade names or commercial products does not constitute en-dorsement or recommendation for use.

The authors are gateful for the support of Lockheed Engineering and Management Services Company, Computer Sciences Corporation and the University of Nevada-Las Vegas, as well as for the valuable services of Mr. Boyd Fagan of Life Systems, Inc.

REFERENCES

[1]   Kovell, S. P., 1986, "Evaluating and Validating Analytical Methods
      Using the Environmental Protection Agency's Contract Laboratory
      Program," Quality Control in Remedial Site Investigation:
      Hazardous and Industrial Solid Waste Testing, Fifth Volume, ASTM
      STP 925, C. L. Perkett, Ed., American Society for Testing and
      Materials, Philadelphia, Pennsylvania, 1986.
[2]   Booth, R. L. and Schweitzer, G. E., 1982, EPA quality assurance
      reference materials project. Memorandum of Understanding, Envi-
      ronmental Monitoring and Support Laboratory-Cincinnati and Environ-
      mental Monitoring Systems Laboratory-Las Vegas.
[3]   IFB WA-85-J680, Chemical Analytical Services for Multi-Media
      Multi-Concentration Organics GC/MS Techniques, U.S. EPA,
      Washington, D.C., August 1985.
[4]   IFB WA-85-J091/092, Inorganic Analysis, Multi-Media, Multi-
      Concentration, U.S. EPA, Washington, D.C., July 1985.
[5]   IFB WA-84-A002, Dioxin Analysis, Soil/Sediment Matrix, Multi-
      Concentration, U.S. EPA, Washington, D.C., September 1983.
[6]   Baughman, K. and Kumar, S., Organic Audit Standard Operating
      Procedure, 85-04, Lockheed Engineering and Management Services
      Company, Las Vegas, Nevada, September 1985.
[7]   Fowler, J. and Lee, J., Inorganic Audit Standard Operating Pro-
      cedure, 85-07, Lockheed Engineering and Management Services
      Company, Las Vegas, Nevada, June 1985.
[8]   Kumar, S., Dioxin Audit Standard Operating Procedure, 85-10,
      Lockheed Engineering and Management Services Company, Las Vegas,
      Nevada, September 1985.
[9]   Brilis, G. and Moore, J., Post-Award On-Site Laboratory Evaluations
      for the Contract Laboratory Program, Standard Operating Pro-
      cedure QAD-QA-L-1, EMSL-LV, Las Vegas, Nevada, April 1984.
[10] U.S. Department of Defense, 1963, Military Standard MIL STD 105D,
      Sampling Procedures and Tables for Inspection by Attributes.
[11] Garner, F. C., 1983, Interim Report—A comprehensive scheme for
      auditing contract laboratory data, Lockheed Engineering and
      Management Services Company, Las Vegas, Nevada.

Debra K. White

INORGANIC ANALYTICAL METHODS - GENERAL DESCRIPTION AND QUALITY CONTROL CONSIDERATIONS

REFERENCE: White, D.K., "Inorganic Analytical Methods - General Description and Quality Control Considerations," Quality Control in Remedial Site Investigation: Hazardous and Industrial Solid Waste Testing, Fifth Volume, ASTM STP 925, C.L. Perket, Ed., American Society for Testing and Materials, 1986.

ABSTRACT: The Environmental Protection Agency's (EPA) Contract Laboratory Program (CLP) provides a variety of state-of-the-art chemical analysis services in support of the Superfund program. The CLP utilizes inductively coupled plasma (ICP), flame atomic absorption (FAA), graphite furnace atomic absorption (GFAA), and cold vapor techniques to analyze water and soil/sediment samples for inorganic priority pollutant constituents. The data generated through the sample analyses is used by EPA to determine the existence and potential threat posed to the public and the environment by identified hazardous waste sites. The extensive quality assurance/quality control (QA/QC) system ensures that data generated by the CLP laboratories is of known quality and is legally defensible in Agency enforcement cases.

KEY WORDS: Inductively coupled plasma (ICP), flame atomic absorption (FAA), graphite furnace atomic absorption (GFAA), Contract Laboratory Program (CLP), quality assurance/quality control (QA/QC).

INTRODUCTION AND BACKGROUND

The Contract Laboratory Program (CLP) supports the Environmental Protection Agency's (EPA) Superfund effort under the 1980 Comprehensive Environmental Response, Compensation, and Liability Act (CERCLA) by providing a variety of state-of-the-art chemical analysis services of

Debra K. White is a Project Officer/Chemist with the Environmental Protection Agency, Office of Solid Waste and Emergency Response, Office of Emergency and Remedial Response, Hazardous Response Support Division, Analytical Support Branch, 401 M Street, S.W., Washington, DC 20460.

known quality on a high volume, cost-effective basis. The standardized analytical services offered by the CLP support a variety of Superfund activities, from those associated with the smallest preliminary site investigation to those of large-scale, complex remedial, monitoring and enforcement actions.

The data obtained under these contracts is being used by EPA to determine the existence and potential threat posed to the public and environment by hazardous waste disposal sites. Because the data may be used in civil and/or criminal litigation, the contracts require strict adherence to the specified analytical methodology, documentation requirements and quality assurance procedures.

The CLP's Routine Analytical Services (RAS) contracts provide for the analysis of a standard list of 23 elements and cyanide. The inorganic RAS methods apply to the analysis of water and soil/sediment samples with concentration levels for the inorganic priority pollutant constituents ranging from low or background levels, to medium levels, where an element may comprise up to 15 percent of the total sample.

The inorganic analytical methods currently utilized by the CLP have been in a state of evolution since the start of the program. Methods continuously undergo review, revision and validation. Advances in technology and new methodology are addressed through the forum of regularly held caucuses which are comprised of both EPA and contract laboratory chemists.

This paper will provide an overview of the analytical methods and the quality assurance/quality control (QA/QC) requirements which constitute the standardized RAS procedures for the analysis of Superfund samples for inorganic constituents. Detailed protocols are contained in the most recent version of the CLP Inorganics Statement of Work (1). Detailed information on method performance and quality control considerations is available in the paper "Inorganic Methods - Performance and Quality Control Considerations," included in this volume (2).

ANALYTICAL METHODS

Due to the requirements of the Superfund program for efficient and accurate analyses on a high volume basis, as well as for legally defensible data, it was necessary to develop standardized analytical methods that were applicable to a wide variety of analytes and matrices and which also incorporated detailed QA/QC and documentation requirements. Contract laboratories in the CLP are required to follow these specific analytical methods and QA/QC and documentation procedures in order to comply with the terms of their contracts. While the requirement for strict adherence to the contract protocols may limit flexibility in terms of achieving optimum conditions for the analysis of individual parameters in unusual matrices, it does afford the program the benefits of consistency and efficiency.

Sample Preparation

Acid digestion procedures are employed in the preparation of sediments, sludges, soils and aqueous samples for elemental analysis by inductively coupled plasma (ICP), flame atomic absorption (FAA) and graphite furnace atomic absorption (GFAA) techniques. Separate digestion procedures are employed for the preparation of samples for analysis by ICP/FAA and GFAA, as well as for aqueous and solid samples. The digestion procedures are summarized in Figures 1 and 2. For solid samples, separate aliquots are used to perform a percent solids determination. The final results of all the analyses are then converted to and reported on a dry weight basis.

Sample Analysis

Graphite furnace atomic absorption (GFAA) spectrophotometry methods are contained within the contract's Statement of Work (SOW) for the analysis of antimony, arsenic, beryllium, cadmium, chromium, lead, selenium, silver and thallium. Flame AA methods are given for calcium, magnesium, potassium and sodium. Flameless AA methods (cold vapor) are given for the analysis of mercury. The ICP method contained in the SOW can be utilized for any of the above analytes as long as the contract required detection limit (CRDL) criteria (Table 1) can be met.

TABLE 1 -- Inorganic Constituents -- RAS

ANALYTE	CRDL (ug/L)	ANALYTE	CRDL (ug/L)
Aluminum	200	Manganese	15
Antimony	60	Mercury	0.2
Arsenic	10	Nickel	40
Barium	200	Potassium	5000
Beryllium	5	Selenium	5
Cadmium	5	Silver	10
Calcium	5000	Sodium	5000
Chromium	10	Thallium	10
Cobalt	50	Vanadium	50
Copper	25	Zinc	20
Iron	100		
Lead	5		
Magnesium	5000	Cyanide	10

The ICP method stipulated within the SOW is a modified version of EPA Method 200.7. This method describes a technique for the simultaneous or sequential multi-element determination of trace elements in solution. The basis of the method is the measurement of atomic emission by an optical spectroscopic technique. Background correction techniques are required to compensate for variable background contribution to the determination of the analytes. Method 200.7 includes a discussion of the various types of interferences encountered in ICP analysis (i.e., spectral, physical and chemical). Methods for confirmation of interferences and corrective actions are also delineated.

The atomic absorption methods included in the SOW were derived from EPA's "Methods for Chemical Analysis of Water and Wastes."

FIGURE 1 -- Digestion Procedures for
Aqueous Samples

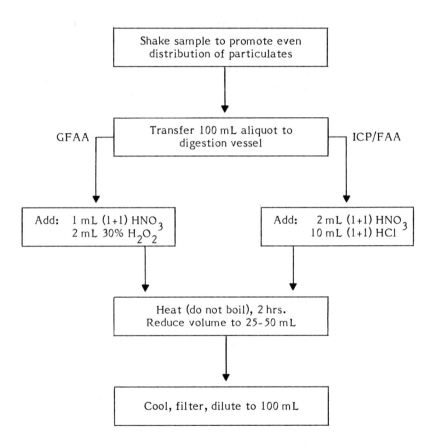

Results reported as "total".

FIGURE 2 -- Digestion Procedures for
Solid Samples

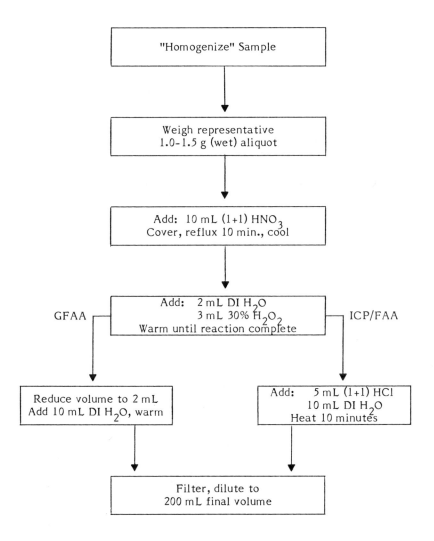

Included within the methods are recommendations for instrument parameters such as wavelength, temperature programs, gases and flame conditions, as well as discussions of potential interferences, matrix modifiers and sources of contamination. In order to accommodate advanced techniques such as the use of L'Vov platforms, the use of the recommended instrument settings is left optional. The application of background correction is required for all GFAA analyses and all flame analyses conducted at wavelengths less than 350 nm. Additionally, for GFAA analyses, the laboratory must demonstrate, on a sample-by-sample basis, that full methods of standard addition (MSA) is not required by performing a single point spike on each sample.

For the analysis of cyanide, three methods, each of which affords a different working range, are allowed: titrimetric, manual spectrophotometric and semiautomated spectrophotometric. The basis of all three methods is the release of HCN from cyanide complexes by means of a reflux-distillation operation, which is then absorbed in a scrubber containing NaOH solution. The cyanide ion is then determined either colorimetrically or by volumetric titration.

Manual and automatic cold vapor techniques for the determination of mercury in both aqueous and solid samples are included in the SOW. In these methods, following acid treatment of the samples, potassium permanganate and potassium persulfate are utilized to oxidize organomercury compounds to the mercuric ion prior to measurement via a flameless AA procedure.

QUALITY ASSURANCE/QUALITY CONTROL

The main objective of the rigorous QA/QC requirements incorporated in the SOW is to assure the production of data of known quality which will meet the client's needs whether it be to support a remedial decision or an enforcement action. The high level of documentation required in conjunction with the built-in QA/QC serves to provide EPA with legally defensible data. The QA/QC aspects contained within the SOW may be divided into two categories: 1) instrument or system performance QA/QC and 2) method performance QA/QC.

Instrument Performance QA/QC

o   Instrument Calibration — The objective in establishing compliance requirements for satisfactory instrument calibration is to insure that the instrument is capable of producing acceptable quantitative data. The contract specifies that instruments must be calibrated daily and each time the instrument is set up. The number of calibration standards are specified for each analytical method. Calibration standards must be prepared using the same type of acid or combination of acids and at the same concentration as will result in the samples following preparation. In addition, for AA analyses, the laboratory is required to analyze a standard at the CRDL to verify their capability to meet the required levels of detection (Table 1).

o   Initial Calibration Verification — Initial calibration verification (ICV) is performed immediately following instrument calibration by analyzing an independent, certified solution to verify and document the accuracy of the initial calibration for each analyte. When ICV results exceed the specified control limits (Table 2), the contract requires that the analysis must be terminated, the problem corrected, the instrument recalibrated and the calibration reverified.

TABLE 2 -- Initial and Continuing Calibration
Control Limits for Inorganic Analyses

ANALYTICAL METHOD	INORGANIC SPECIES	% OF TRUE VALUE	
		LOW	HIGH
ICP/AA	Metals	90	110
Cold Vapor	Mercury	80	120
Other	Cyanide	85	115

o Calibration Blank — A calibration blank (CB) is analyzed each time the instrument is calibrated, at the beginning and end of the run, and at a frequency of 10% during the run. The CB serves as a monitor of instrument drift and possible memory effects or contamination. If the CB result is greater than the CRDL, the contract requires the analysis be terminated, the problem identified and corrected, and the instrument recalibrated.

o Continuing Calibration Verification — To assure calibration accuracy and monitor instrument performance during each analysis run, a continuing calibration verification (CCV) standard is analyzed at a frequency of 10% or every two hours, whichever is more frequent. The analyte concentration in the CCV standard must be at or near the mid-range levels of the calibration curve. The control limits and corrective action procedures as discussed under initial calibration verification apply as well for CCV.

o ICP Interference Check Sample Analysis — The ICP Interference Check Sample (ICS) analysis is performed to verify the ICP's interelement and background correction factors. An EPA supplied ICS must be run at the beginning and end of each analysis run or at a minimum of twice per eight-hour working shift, whichever is more frequent. If the results of the ICS do not meet the contract specified control limit of $\pm 20\%$ of the true value, the analysis must be terminated, the problem corrected, the instrument recalibrated and the samples reanalyzed.

o ICP Linear Range Analysis — To verify linearity near the CRDL, the contractor must analyze an ICP standard at a concentration level of two times the CRDL. This standard must be run at the beginning and end of each sample analysis run or at a minimum of twice per eight-hour working shift, whichever is more frequent. The upper limit of the ICP linear range must be verified on a quarterly basis. The analytically determined concentration of the standard used to define this upper limit must fall within $\pm 5\%$ of the true value. This concentration defines the limit beyond which results cannot be reported under the contract without dilution into the working range.

o   Instrument Detection Limits — Instrument detection limits (IDL) must be determined by multiplying by three, the average of the standard deviations obtained on three nonconsecutive days from the analysis of a standard solution (each analyte in reagent water) at a concentration 3-5 times the IDL, with seven consecutive measurements per day. Updated IDLs are reported on a quarterly basis.

Method Performance QA/QC

o   Preparation Blanks — The analysis of preparation blanks provides EPA a means of assessing the existence and magnitude of contamination introduced via the analytical scheme. At least one preparation blank, consisting of deionized distilled water processed through each sample preparation procedure performed, must be prepared and analyzed with each batch of samples digested or for each 20 samples received, whichever is more frequent. If the concentration of the blank is above the CRDL, then any samples with less than 10 times the concentration level identified in the blank must be redigested and reanalyzed. The reported sample results are not corrected for the blank results.

o   ICP Serial Dilution Analysis — Serial dilution analysis is performed to ascertain whether significant physical or chemical interferences exist due to the sample matrix. One sample from each group of samples of a similar matrix type and concentration from each Case (group of samples from a given site over a given time period) must undergo at least one serial dilution analysis. If the analyte concentration is minimally a factor of 10 above the IDL after dilution, the analysis must agree within 10% of the original determination. If the result is not within 10%, a chemical or physical interference effect should be suspected and the data flagged.

o   Laboratory Control Sample Analysis — The laboratory control sample (LCS) analysis is designed to serve as a monitor of the efficiency of the digestion procedure. An aqueous LCS must be analyzed on a digestion batch basis, whereas a solid LCS is analyzed on a monthly basis. Both the aqueous LCS and the solid LCS are provided by EPA to the contract laboratories. Results of the LCS analyses must fall within the EPA established control limits. Results which fall outside the specified control limits are indicative of an analytical problem related to the digestion/sample preparation procedures and/or instrument operations.

o   Spiked Sample Analysis — Spiked sample analysis is designed to provide information about the effect of the sample matrix on the digestion and measurement methodology. The spike is added prior to digestion/distillation steps. Spiked sample analyses are performed for each matrix and concentration level classification in each Case of samples. Analyte spiking levels are specified within the contract. A variety of factors can impact the outcome of the spike sample results; these include matrix suppression or enhancement effects, duplicate precision, digestion efficiency, contamination and the relative levels of analyte in the sample and the spike. If the spike recovery falls outside the limits of 75-125%, the data associated with that spiked sample must be flagged.

o Duplicate Sample Analysis — Duplicate sample analysis is performed for each matrix and concentration level classification within each Case. The results of the duplicate analyses serve as an indicator of the precision of the method and the sample results. The non-homogeneous nature of typical Superfund samples has made it impractical to define contractually required control limits. As it is difficult to determine whether poor precision is a result of sample non-homogeneity, method inadequacies or laboratory technique, the control limits given in the Statement of Work are to be used for flagging data rather than for taking corrective actions.

o Furnace Atomic Absorption QC Analysis — Duplicate injections and post digestion analytical spikes are incorporated into the QC scheme in order to establish a mechanism whereby the data user can estimate the precision and accuracy of the individual analytical determination relative to the overall method precision and accuracy. The post digestion spike result is used to determine whether full MSA is required for quantitation.

CONCLUSION

The use of standardized analytical methodology which incorporates rigorous QA/QC and documentation requirements, has served to consistently provide the Agency's Superfund effort with data of known quality. The CLP is constantly striving to improve the quality of the data produced, by continually reviewing, updating and refining the current methods to accommodate advanced technology and new methods based upon program needs.

ACKNOWLEDGEMENT

The author acknowledges with appreciation the assistance of Steven Manzo, Sample Management Office, in the preparation of this paper.

REFERENCES

(1)   IFB WA 85-J838/J839, Inorganic Analysis, Multi-Media, Multi-Concentration, USEPA, Washington, DC, July 1985.

(2)   Aleckson, Keith A., Fowler, John W., and Lee, Y. Joyce, "Inorganic Methods - Performance and Quality Control Considerations," Quality Control in Remedial Site Investigation: Hazardous and Industrial Solid Waste Testing, Fifth Volume, ASTM STP 925, C.L. Perket, Ed., American Society for Testing and Materials, 1986.

Keith A. Aleckson, John W. Fowler, and Y. Joyce Lee

INORGANIC ANALYTICAL METHODS PERFORMANCE AND QUALITY CONTROL
CONSIDERATIONS

REFERENCE: Aleckson, Keith A., Fowler, John W., and Lee, Y.
Joyce, "Inorganic Analytical Methods Performance and Quality
Control Considerations," Quality Control in Remedial Site
Investigation: Hazardous and Industrial Solid Waste Testing,
Fifth Volume, ASTM STP 925, C. L. Perket, Ed., American Society
for Testing and Materials, Philadelphia, 1986.

ABSTRACT: The United States Environmental Protection
Agency's Contract Laboratory Program was established to
produce analytical services in support of Superfund
(CERCLA) site investigations and to provide legally defensible analytical results of known and documented data
quality. The inorganic analytical protocol was developed
for the analysis of 24 metals in aqueous and solid matrices.
Methods are included in the protocol for inductively coupled
argon plasma, heated graphite furnace atomic absorption, and
mercury cold vapor spectroscopy. Precision, accuracy, detection limits, and other performance criteria are monitored
by the Environmental Monitoring Systems Laboratory - Las
Vegas. The observed performance of this protocol in the
production-line analysis of thousands of samples by contractor laboratories is presented in this paper.

KEYWORDS: CERCLA, inorganic analytical methods, Superfund,
contract laboratory program, inductively coupled argon
plasma, heated graphite furnace atomic absorption, mercury
cold vapor, precision, accuracy, detection limit, protocol,
water matrix, soil matrix.

The Environmental Protection Agency's Environmental Monitoring
Systems Laboratory - Las Vegas (EMSL-LV) is responsible for conducting

Keith A. Aleckson is a Scientist and Y. Joyce Lee is an Associate
Scientist at Lockheed Engineering and Management Services Co., P.O.
Box 15027, Las Vegas, NV 89114. Mr. Fowler is a Chemist in the Toxic
and Hazardous Waste Operations Branch of the USEPA Quality Assurance
Division at the Environmental Monitoring Systems Laboratory, 944 East
Harmon, Las Vegas, NV 89109.

a quality assurance program in support of the Agency's Superfund Contract Laboratory (CLP) program. Information on the CLP and the EMSL-LV quality assurance support program is presented in the papers by Kovell [1] and Moore and Pearson [2] within this volume. Method performance and laboratory performance data on inorganic methods from 18 commercial laboratories using the Superfund inorganic analytical protocol [3] are monitored by EMSL-LV through: 1) a computer quality assurance/quality control (QA/QC) data base, 2) a performance evaluation (PE) sample program, and 3) laboratory QC materials and standards prepared at the EMSL-LV. The EMSL-LV continuously monitors individual laboratory performance and performs dynamic validation of the analytical methods, i.e., the performance of the methods is continuously assessed and periodically documented. If these studies indicate that method performance does not meet the Agency's monitoring goals, studies are conducted to gather the data necessary to improve method performance.

The Superfund inorganic analytical protocol contains methods and procedures for the analysis of 24 metals and cyanide in solid and aqueous matrices at concentration levels from parts-per-billion up to 15 percent of the sample weight. This version of the protocol has been used since October 1984. Additional information on the inorganic analytical protocol is available in the paper presented by White [4] within this volume. The objective of this paper is to document the performance of the Superfund inorganic analytical protocol for aqueous and solid matrices and to compare this performance to objectives for precision and accuracy on field samples [5].

PROTOCOL DESCRIPTION

The Superfund inorganic analytical protocol contains sample digestion procedures, instrumental analysis methods, and QA/QC requirements. Sample digestion is required for all samples of both aqueous and solid matrices. Separate digestion procedures exist for inductively coupled argon plasma (ICP), graphite furnace atomic absorption (FAA), and mercury cold vapor (CV) analysis. All digestions, except mercury, are filtered before analysis. The digestion procedure followed depends on the metal to be analyzed and the instrumental method. Table 1 summarizes the different digestion procedures used in the protocol.

Only results from the techniques listed in Table 1 are presented in this paper.

The following QA/QC checks are required by the protocol:

1. Initial Calibration and Calibration Verification--Instrument calibration is checked with an independent standard before sample analysis.
2. Continuing Calibration Verification--Continuing instrument calibration is checked during the analysis run after every 10 samples.
3. Preparation Blank Analysis--A blank is carried through the entire preparation with each digestion batch for every matrix and analyzed to check for laboratory contamination.

TABLE 1--Digestion procedures.

Reagents	Method of Analysis	Element(s)
HNO3/HCl/H2O2	ICP[a]	Al, Sb, Ba, Be, Ca, Cd, Cr, Co, Cu, Fe, Pb, Mg, Mn, Ni, K, Ag, Na, Tl, Sn, V, and Zn.
HNO3/HCl/H2O2	FAA[b]	Sn and Sb.
HNO3/H2O2	FAA	As, Be, Cd, Cr, Co, Cu, Fe, Pb, Mn, Ni, Se, Ag, Tl, V, and Zn.
H2SO4/HNO3/Permanganate	CV[c]	Hg

[a] ICP - Inductively Coupled Argon Plasma
[b] FAA - Furnace Atomic Absorption
[c] CV - Mercury Cold Vapor

4. ICP Interference Check Analysis--Analysis of an ICP interference check solution is required before and after sample analyses by
ICP or every eight hours, whichever is more frequent. The ICP Interference Check contains high levels of interference metals and low
levels of analyte metals.
5. ICP Serial Dilution--ICP serial dilution is required as a check
for sample matrix effects on one sample of each matrix or group of 20
samples, whichever is more frequent.
6. Pre-digestion Matrix Spike Analysis--One sample of every 20
samples in each matrix is spiked before digestion to determine accuracy
of recovery on that matrix.
7. Duplicate Sample Analysis--One sample of every 20 samples in
each matrix is digested and analyzed in duplicate to determine precision on that matrix.
8. Furnace AA QC Analysis--Dilution, analytical spike, and method
of standard addition are performed as required on each sample.
9. Aqueous Laboratory Quality Control Sample Analysis--Standards
are carried through the entire digestion and analysis procedures to
test laboratory performance with every sample batch.
10. Solid Laboratory Quality Control Sample Analysis--A solid
reference material is carried through the protocol once per month to
test laboratory performance.
11. Quarterly Determination of Instrument Detection Limits and ICP
Linear Ranges--Instrument detection limits (IDL's) and ICP linear
ranges are determined quarterly by protocol specified methods.
12. ICP Interelement Correction Factors--ICP interelement correction factors are reported quarterly but the protocol does not specify
how they are determined.

Steps 1 through 9 are carried out on every batch of samples within the
Contract Laboratory Program (CLP) and reported along with sample results on standardized QC forms. The data on the QC forms is entered
into the QA/QC data base maintained by the EMSL-LV. A batch of samples
is defined as twenty samples or less, which are of the same matrix, and
are digested at the same time. Several batches may make up one case.

A quarterly performance evaluation (PE) sample program is conducted by the EMSL-LV. PE samples are sent out either as single blind,
recognizable PE samples, or as double blind, unrecognizable PE samples.
Both aqueous and solid matrix PE samples are routinely sent each quarter. PE samples are prepared at the EMSL-LV or obtained from an outside source such as National Bureau of Standards (NBS). Table 2 is

TABLE 2--PE sample and LCS concentrations.

Ele-ment	PE 1 Aqueous	PE 1 Solid	PE 2 Aqueous	PE 2 Solid	PE 3 Aqueous	PE 3 Solid	PE 4 Solid	LCS Solid
Al	--	6595	3000	2	1000	10553	22600	4560
Sb	180	--	600	--	420	--	(51)	--
As	50	--	150	--	100	10.5	(66)	17.0
Ba	800	--	1500	246	1200	10.4	--	--
Be	30	--	40	--	45	19.5	--	--
Cd	25	5.5	50	--	35	20.0	10.2	19.1
Ca	1000	--	10000	4127	30000	2664	(29000)	--
Cr	50	8.5	100	65	150	46.5	29600	193
Co	600	--	1000	--	200	71.5	--	--
Cu	125	--	250	33	175	39.0	109	1080
Fe	--	5028	200	31109	800	14350	113000	16500
Pb	--	11.5	--	--	30	55.0	714	526
Mg	10000	--	10000	7799	40000	2428	7400	--
Mn	30	73.5	150	602	150	169	785	202
Hg	5	--	--	--	20	26.5	1.1	16.3
Ni	--	--	--	67	160	44.0	45.8	194
K	10000	--	10000	--	20000	--	12600	--
Se	50	--	--	--	50	13.5	(1.5)	--
Ag	--	--	--	--	--	--	--	50.6
Na	10000	--	10000	--	45000	--	5400	--
Tl	100	--	80	--	100	15.0	1.44	--
Sn	--	--	--	--	160	--	--	--
V	200	--	70	63	150	60.0	23.5	13.0
Zn	150	19.0	50	57	800	63.5	1720	1320

Water results in µg/L, Theoretical Values.
Solid results in mg/kg, Mean CLP analytical values.
-- Not present, below detection limit, or not determined.
 PE 1, PE 2, and PE 3 "waters" are synthetic samples and PE 1, PE 2,
 and PE 3 "solids" are spiked soils.
 PE 4 is the NBS SRM 1645, River Sediment.
 LCS Solid is the Dried Municipal Sludge Material prepared by
 the Environmental Monitoring and Support Laboratory,
 Cincinnati, Ohio (EMSL-Cincinnati).
( ) Uncertified value.

a list of concentration levels of PE samples and the solid Laboratory
Control Sample (LCS) which have been analyzed under the Superfund
inorganic analytical protocol.

Theoretical values of water PE samples as well as laboratory results of
the solid LCS and solid PE samples are used to determine precision and
accuracy.

METHOD PERFORMANCE RESULTS

 Method performance on field sample analyses is monitored through
the QA/QC data base. An outlier test has been applied to matrix spike
data on field samples and the LCS recovery values. Results outside of
three standard deviations from the mean have been excluded from the
QA/QC data base.

Grubbs' test [6] was applied on PE samples to eliminate outliers. This test was used because of the smaller number of points available in the PE sample studies. Points outside of the 5% significance level were excluded from precision and accuracy calculations.

Precision

Precision expresses the reproducibility of an analytical method, both the error associated with the sample preparation and the instrumental analysis. Intra-laboratory precision, precision within a single laboratory, and inter-laboratory precision, precision between laboratories, are presented in Table 3. This table summarizes 40 to 50 data points obtained over six months, generated by 12 laboratories' analysis of the solid LCS. Results of the aqueous LCS are not included because a common solution was not used as the aqueous LCS during this period. Percent relative standard deviation (%RSD) is used as a measure of precision. Percent RSD is defined as the standard deviation divided by the mean value, multiplied by 100 to convert into percent.

TABLE 3--Solid LCS precision.

Element	Method	Pooled Intra-Lab %RSD	Total Inter-Lab %RSD
Al	ICP	15.2	16.1
Cd	ICP	13.9	15.7
Cr	ICP	10.0	12.0
Cu	ICP	8.4	9.3
Fe	ICP	8.8	10.2
Pb	ICP	9.9	10.9
Mn	ICP	6.4	13.1
Hg	CV	42.3	48.3
Ni	ICP	13.3	14.9
Ag	ICP	20.2	23.9
Zn	ICP	15.1	19.8

Only those elements which are present in the solid LCS at levels above contract required detection limits are listed. Elements at or below the detection limit show high %RSD values due to the dominance of instrumental noise over analyte signal at very low levels.

A new solid LCS has been developed by the EMSL-LV through the University of Nevada, Las Vegas Quality Assurance Laboratory. The new LCS has replaced the Dried Municipal Sludge material from EMSL-Cincinnati. A greater range of elements and greater degree of homogeneity are present in the new LCS material.

Method inter-laboratory precision is also determined through the analysis of PE samples. Inter-laboratory PE sample results are presented in Table 4 for both aqueous and solid matrices.

Field sample inter-laboratory precision is estimated through analysis of duplicate samples from every matrix associated with a batch of samples. Median %RSD values from the QA/QC data base for every element

TABLE 4--Inter-laboratory performance evaluation sample precision.

		PE 1		PE 2		PE 3		PE 4
		%RSD		%RSD		%RSD		%RSD
Element	Method	Water	Solid	Water	Solid	Water	Solid	Solid
Al	ICP	--	41.8	6.83	22.2	11.4	15.4	14.4
Sb	ICP	14.5	52.5	7.54	--	10.5	--	--
As	FAA	4.14	28.6	10.0	22.8	14.1	--	--
Ba	ICP	4.75	13.8	10.0	12.2	5.79	12.8	--
Be	ICP	7.54	--	28.6	--	10.2	13.7	--
Cd	ICP	16.8	7.10	6.12	--	13.8	17.6	33.3
Ca	ICP	4.54	46.7	4.50	7.50	8.88	32.1	--
Cr	ICP	13.6	35.4	7.00	13.8	8.83	12.1	7.8
Co	ICP	4.60	--	5.50	--	10.0	10.6	--
Cu	ICP	9.60	21.1	4.06	12.1	6.57	12.2	11.2
Fe	ICP	--	23.9	14.7	8.30	6.11	16.9	10.7
Pb	FAA	--	29.6	--	--	32.2	--	9.2
Mg	ICP	6.58	17.9	4.72	18.6	8.40	13.7	7.5
Mn	ICP	6.07	9.10	5.30	10.0	7.08	19.2	9.4
Hg	CV	17.0	--	--	--	20.5	38.1	25.0
Ni	ICP	--	--	--	7.50	9.04	12.9	15.0
K *	ICP	20.9	47.5	13.9	--	13.8	--	--
Se	FAA	8.00	--	--	--	18.0	18.4	--
Ag	ICP	--	--	--	--	--	--	--
Na	ICP	8.66	17.4	11.9	--	5.49	--	--
Tl	FAA	--	--	25.8	--	9.68	24.9	--
Sn	ICP	--	--	--	--	--	--	44.1
V	ICP	12.2	--	5.80	19.0	4.85	23.6	--
Zn	ICP	8.10	13.9	12.7	14.0	6.36	12.3	5.8

-- Not present or below detection limit.
* A small amount of Flame Atomic Absorption data is included in the
potassium results.

are presented in Table 5. Only duplicate analysis where the sample concentration is over five times the contract required detection limit are
counted in determining the median values. Duplicate results below five
times the contract required detection limits were not used because of
the expected increase in %RSD near the detection limit. The Agency objectives for field sample precision is for RSD values to fall below 14%.

Percent RSD values from Table 4 and Table 5 are not directly comparable. Table 4 shows the %RSD values calculated from multi-laboratory analysis of the same material in each study. Values shown in
Table 5 are the median %RSD for duplicate field sample analysis within
single laboratories. Precision is better as expected in the single
laboratory data.

A soil homogenization procedure will be added to the latest CLP
inorganic analytical protocol through a contract modification. The
results presented in Table 5 were collected before the adoption of the
homogenization procedure.

Method Bias

Method bias is an estimate of the difference between the average

TABLE 5--Duplicate %RSD medians from field samples.

Element	Method	Aqueous	Solid	Element	Method	Aqueous	Solid
Al	ICP	6.14	8.70	Mg	ICP	1.05	8.41
Sb	ICP	4.46	--	Mn	ICP	1.68	8.20
As	FAA	5.6	10.6	Hg	CV	5.16	7.21
Ba	ICP	1.32	7.21	Ni	ICP	2.34	3.09
Be	ICP	1.06	--	K	ICP	1.87	--
Cd	ICP	4.70	13.8	Se	FAA	6.76	9.18
Ca	ICP	1.21	10.5	Ag	ICP	--	--
Cr	ICP	4.95	11.2	Na	ICP	1.58	--
Co	ICP	1.25	--	Tl	FAA	3.14	--
Cu	ICP	1.76	9.26	Sn	ICP	--	--
Fe	ICP	2.64	8.06	V	ICP	6.18	10.8
Pb	FAA	5.83	10.8	Zn	ICP	2.21	7.00

-- Insufficient number of data points to determine %RSD median (N <20)

value of a set of measurements of a sample of known concentration and the true value of the sample. Bias can be caused by the sample matrix, sample preparation procedure, analytical method, and/or by improper laboratory bench procedures. The percent bias is defined as:

$$\% \ B = 100 \times (\overline{X} - T) / T \qquad (1)$$

where

B = Bias,
T = True or reference value,
$\overline{X}$ = Mean of the measured values.

Equation (1) results in a negative bias value when the mean recovery is less than complete. Positive bias values result when the methods and procedures being tested yield average results above the theoretical or certified values.

Method bias in the inorganic analytical protocol is determined through the analysis of LCS and PE samples. Bias resulting from sample matrix is determined through the matrix spike analysis. Matrix spikes are required on one sample of every matrix contained within a batch of samples.

Method bias measured by the analysis of aqueous and solid LCS's is presented in Table 6. Aqueous LCS's are mid-range standards which are digested and analyzed with each group of samples. The solid LCS is the Dried Municipal Sludge material from EMSL-Cincinnati which is digested and analyzed once per month.

Between 300 and 600 points for each element where used to calculate the percent bias on the aqueous LCS. Approximately forty (40) points were available for calculation of bias on the solid LCS. Far fewer points are available on the solid LCS because it is only analyzed monthly.

Perhaps a more realistic estimate of method bias is available through the PE sample program. These samples are of unknown concentration to the laboratories conducting the analysis. In the case of the

TABLE 6--Method % bias estimated from laboratory control samples.

Element	Method	Aqueous LCS	Solid LCS	Element	Method	Aqueous LCS	Solid LCS
Al	ICP	4.0	-10.6	Mg	ICP	-0.8	--
Sb	ICP	-2.3	--	Mn	ICP	-1.1	1.0
As	FAA	-3.0	-33.0	Hg	CV	-1.4	-26.4
Ba	ICP	0.0	--	Ni	ICP	-0.1	-11.0
Be	ICP	-1.5	84.0	K	ICP	-1.9	--
Cd	ICP	-3.4	-2.9	Se	FAA	-2.7	--
Ca	ICP	2.0	--	Ag	ICP	-12.2	-14.4
Cr	ICP	-1.4	-7.3	Na	ICP	0.0	--
Co	ICP	-3.3	--	Tl	FAA	-2.9	--
Cu	ICP	0.0	-2.6	Sn	ICP	-2.4	--
Fe	ICP	3.0	3.0	V	ICP	-1.2	-7.6
Pb	FAA	-1.4	--	Zn	ICP	4.0	-4.3

-- Not present or below detection limit.

aqueous PE samples, the samples are sent double blind to the laboratories. Unrecognizable PE samples may be treated in a more routine manner than recognizable PE samples. Bias estimates obtained from the PE program are shown in Table 7.

Aqueous PE samples one through three are synthetically prepared spikes of reagent water. The solid PE 4 is the NBS River Sediment, SRM 1645. Theoretical values were used to calculate the bias on the aqueous samples, while the certified values were used to calculate bias on the NBS River Sediment. Low recovery values on the NBS River Sediment may be due to the more rigorous sample preparation and total analytical methods used to set the certified concentration on this material.

Field sample accuracy is estimated through the analysis of pre-digestion matrix spikes on one sample of every matrix associated with a batch of samples. Mean percent bias results for field samples are presented in Table 8. The Agency objective for accuracy on field samples is to achieve a bias of not over 25%, i.e., spike recovery of between 75% and 125% of the amount added to the sample.

Detection Limits

A requirement of the inorganic protocol is the quarterly determination of instrument detection limits (IDL's). An IDL is defined as the minimum concentration of an element that can be identified in a clean water matrix and reported with a 99% confidence that the metal is present in the solution. A method of determining IDL's from replicate analyses, based on the work of Budde [7], is described within the protocol. Three sets of seven measurements of a standard at three to five times the estimated detection limit are taken on non-consecutive days. These measurements are used to calculate a standard deviation for each element. The IDL's are set at three times the standard deviation for each element.

TABLE 7--Method % bias estimated from PE samples.

Element	Method	PE 1 Aqueous	PE 2 Aqueous	PE 3 Aqueous	PE 4 Solid
Al	ICP	--	-7.3	-1.2	-78.8
Sb	ICP	-13.9	-5.0	-8.8	--
As	FAA	-3.8	-6.7	-14.4	--
Ba	ICP	-2.1	-3.0	-6.7	--
Be	ICP	6.0	5.0	0.0	--
Cd	ICP	-2.4	-2.0	-5.4	2.9
Ca	ICP	-0.4	0.0	-4.3	-4.2
Cr	ICP	-4.6	0.0	-3.3	-6.1
Co	ICP	-0.3	-1.9	-6.5	--
Cu	ICP	0.0	-1.6	-1.7	-2.5
Fe	ICP	--	9.0	4.0	-27.0
Pb	FAA	--	--	-0.7	-2.2
Mg	ICP	0.0	-3.5	-3.9	-10.6
Mn	ICP	0.0	1.0	-4.0	-15.1
Hg	CV	-2.2	--	-26.5	-9.1
Ni	ICP	--	--	-2.5	-17.0
K	ICP	-11.9	-13.3	-11.0	--
Se	FAA	-3.4	--	-13.6	--
Ag	ICP	--	--	--	--
Na	ICP	-0.4	-0.8	-7.3	--
Tl	FAA	2.0	-17.5	3.0	--
Sn	ICP	--	--	-2.5	--
V	ICP	4.0	-1.4	-4.0	--
Zn	ICP	2.0	10.0	-2.9	-6.2

-- Not present, below detection limit, or no accepted value available.

TABLE 8--Method % bias estimated from field sample matrix spikes.

Element	Method	Aqueous	Solid	Element	Method	Aqueous	Solid
Al	ICP	-4.6	3.0	Mg	ICP	-5.8	-8.0
Sb	ICP	-4.0	-33.5	Mn	ICP	-6.6	-3.5
As	FAA	-8.9	-34.8	Hg	CV	-5.9	-3.0
Ba	ICP	-6.6	-3.1	Ni	ICP	-5.8	-2.7
Be	ICP	-1.7	-1.4	K	ICP	-4.4	-19.2
Cd	ICP	-1.5	-1.5	Se	FAA	-19.6	-24.3
Ca	ICP	-5.3	-0.7	Ag	ICP	-19.7	-14.8
Cr	ICP	-5.6	-4.2	Na	ICP	-5.4	-3.1
Co	ICP	-8.1	-4.1	Tl	FAA	-26.4	-9.8
Cu	ICP	-4.6	-5.5	Sn	ICP	-17.1	-19.9
Fe	ICP	-8.3	-8.4	V	ICP	-4.1	-4.4
Pb	FAA	-11.4	-8.2	Zn	ICP	-4.9	-3.6

Bias in Table 8 is based on the recovery of the pre-digested spike.

Contract Required Detection Limits (CRDL's), listed in the protocol, must be met before samples can be analyzed. IDL's must be lower
than or equal to the CRDL's. Meeting the CRDL's requires the use of
furnace atomic absorption methods for several elements.

The IDL's calculated by these procedures are a measure of instrument performance. Actual detection limits on samples may be higher
because of the added variance of the sample preparation procedures and
possible sample matrix effects. Mean IDL's for the most commonly used

analytical wavelengths of the laboratories using this protocol are
reported in Table 9.

Analytical wavelengths selection are left to the choice of the
laboratory. The only requirement on wavelength selection is that the
contract required detection limits be met.

TABLE 9--Wavelength and instrument detection limits.

Element	CRDL	Method	N	IDL Mean	IDL Std Dev	Wave-Length (nm)
Al	200	ICP	7	70.7	59.3	309.3
Sb	60	ICP	5	42.3	11.3	217.6
As	10	FAA	18	4.6	2.3	198.7
Ba	200	ICP	5	22.1	31.7	493.4
Be	5	ICP	10	2.3	1.7	312.0
Cd	5	ICP	5	4.0	1.1	228.8
Ca	5000	ICP	7	529	472	317.9
Cr	10	ICP	9	5.8	2.9	267.7
Co	50	ICP	11	11.4	8.5	228.6
Cu	25	ICP	11	9.7	6.5	324.5
Fe	100	ICP	10	27.4	20.9	259.9
Pb	5	ICP	12	2.3	1.2	283.3
Mg	5000	ICP	11	385	449	279.6
Mn	15	ICP	10	5.2	4.6	257.6
Hg	0.2	CV	12	0.2	0.1	253.7
Ni	40	ICP	9	17.8	10.1	232.0
K	5000	ICP	8	668	444	766.5
Se	5	FAA	18	2.8	1.3	196.0
Ag	10	ICP	10	5.4	2.7	328.1
Na	5000	ICP	9	756	864	589.0
Tl	10	ICP	18	4.3	2.4	276.8
Sn	40	ICP	7	23.8	8.4	190.0
V	50	ICP	10	13.1	10.0	292.5
Zn	20	ICP	0	8.3	6.3	213.9

IDL - Instrument Detection Limit (µg/L).
N - Number of laboratories using the most common wavelength.
CRDL - Contract Required Detection Limit (µg/L).

CONCLUSION

The Superfund inorganic analytical protocol has been used for over
one year by 18 commercial laboratories. During this period, laboratory
and method performance have been monitored by EMSL-LV. Precision for
aqueous samples is in the range of 5-10 %RSD, and 10-30 %RSD on solid
samples. Median %RSD values on aqueous field samples containing over
five times the contract required detection limits range 5-10%, and
5-15% on solid field samples. Method bias for aqueous samples ranges
from 5-10%, while for solid samples the method bias ranges from 10-30%.
Mean bias on field samples ranges from 5-10% for most elements on both
aqueous and solid matrices. The poorest recovery values occur for Sb,
As, and Tl on solid samples and for Tl, Ag, and Se on aqueous samples.

In almost all cases the precision and accuracy is better on those elements analyzed by ICP. Objectives for precision [5] of less than 20% Relative Percent Difference, which equals a RSD of 14%, on field samples with concentrations greater than five times the contract required detection limit and accuracy of less than 25% bias have been met in most instances.

ACKNOWLEDGMENTS

Although the research described in this article has been funded wholly or in part by the United States Environmental Protection Agency through Contract Number 608-03-3249 to Lockheed Engineering and Management Services, Co., it has not been subjected to Agency review and therefore does not necessarily reflect the views of the Agency and no official endorsement should be inferred. Mention of names or commercial products does not constitute endorsement or recommendation for use.

The authors are grateful for the invaluable assistance provided by Charles Upham, Mary Ann Herrington, and Forest Garner in gathering and analyzing contract/laboratory data.

REFERENCES

[1]   Kovell, S. P., 1986, "Evaluating and Validating Analytical Methods Using the Environmental Protection Agency's Contract Laboratory Program," Quality Control in Remedial Site Investigation: Hazardous and Industrial Solid Waste Testing, Fifth Volume, ASTM STP 925, C. L. Perkett, Ed., American Society for Testing and Materials in this publication.

[2]   Moore, J. M. and J. G. Pearson, 1986, "Quality Assurance Support for the Superfund Contract Laboratory Program," Quality Control in Remedial Site Investigation: Hazardous and Industrial Solid Waste Testing, Fifth Volume, ASTM STP 925, C. L. Perkett, Ed., American Society for Testing and Materials in this publication.

[3]   IFB WA-85-J091/J092, Inorganic Analysis, Multi-Media, Multi-Concentration, U.S. EPA, July 1985.

[4]   White, D., 1986, "Inorganic Analytical Methods - General Description and Quality Assurance Considerations," Quality Control in Remedial Site Investigation: Hazardous and Industrial Solid Waste Testing, Fifth Volume, ASTM STP 925, C. L. Perkett, Ed., American Society for Testing and Materials in this publication.

[5]   White, D. and et al., Laboratory Data Validation - Functional Guidelines for Evaluating Inorganic Analysis, United States Environmental Protection Agency, Office of Emergency Remedial Response.

[6]   Grubbs, F. E., "Sample Criteria for Testing Outlying Observations," Annals of Mathematical Statistics, Vol. 21, March 1950, pp. 27-58.

[7]   Glaser J. A., Foerst D. L., McKee G. D., Quave S. E., and Budde, W. L., "Trace Analysis for Wastewaters," Environmental Science and Technology, 1981, Vol. 15, p. 1426.

Robert D. Kleopfer

DIOXIN ANALYTICAL METHODS - GENERAL DESCRIPTION AND QUALITY
CONTROL CONSIDERATIONS

REFERENCE: Kleopfer, R.,"Dioxin Analytical Methods - General
Description and Quality Control Considerations," Quality
Control in Remedial Site Investigation: Hazardous and
Industrial Solid Waste Testing, Fifth Volume, ASTM STP 925,
C. L. Perket, Ed., American Society for Testing and Materials,
Philadelphia, 1986.

ABSTRACT: 2,3,7,8-Tetrachlorodibenzo-p-dioxin is of unique
environmental concern because of its extreme toxicity and
persistency. A well-documented procedure was developed to
measure parts per billion levels in soil using high resolution gas chromatography/low resolution mass spectrometry.
The method utilizes isotope dilution quantitation procedures
with selected ion monitoring. A comprehensive quality assurance plan was developed to evaluate and verify data quality.
Modified methods, based on low resolution mass spectrometry
and tandem mass spectrometry, were developed to provide high
quality data on a more timely basis.

KEYWORDS: 2,3,7,8-TCDD, isotope dilution mass spectrometry,
quality assurance, mass spectrometry

Dr. Kleopfer is Chief of the Contract Laboratory and Quality
Assurance Section at the U.S. Environmental Protection Agency,
Region VII Laboratory, 25 Funston Road, Kansas City, Kansas 66115.

INTRODUCTION

In 1982, numerous sites of potential dioxin (2,3,7,8-tetra-
chloro-dibenzo-p-dioxin; 2,3,7,8-TCDD) contamination were identified
in eastern Missouri. These sites were contaminated in the early
1970s when dioxin-contaminated waste oil was applied to horse
arenas and road surfaces primarily for the purpose of dust control[1].
The chemical waste originated from a plant in Verona, MO, which
produced hexachlorophene, a germicide[2]. The 2,3,7,8-TCDD was a
by-product from the production of 2,4,5trichlorophenol (2,4,5-TCP),
which is used to synthesize hexachlorophene. The public health
implications required a rapid, yet reliable assessment of these
sites. Specifically, a costeffective procedure was required to
analyze for 2,3,7,8-TCDD down to 1 part per billion (ppb) in
hundreds of soil samples.

An analytical procedure was needed that met the following
requirements:

° A method both safe and rugged since several laboratories would be required to support the workload of hundreds of soil samples per week.

° A method providing reasonably rapid results, because of the serious public health implications and intense public interest.

° A relatively simple method readily performed on thousands of samples.

° A method detecting dioxin in soil down to at least 1.0 ppb (micrograms TCDD per kilogram of soil).

° An isomer-specific method, since TCDD toxicity varies greatly from isomer to isomer; the requirement was that 2,3,7,8-TCDD (the most toxic isomer) must be distinguished from the other 21 possible TCDD isomers.

° A reasonably precise and unbiased method compatible with the stated quality assurance objectives[3].

ANALYTICAL METHOD

The requirements, described above, were achieved using a detailed analytical procedure first drafted in December 1982[4]. Subsequent refinements resulted in a document[5] which has been incorporated into a contract for providing analytical services to the U.S. Environmental Protection Agency.

A mixture of 20 ml methanol and 150 ml hexane is used to extract 10 g aliquots of soil that have been thoroughly mixed with 20 g of anhydrous sodium sulfate. This drying step was considered important because many clay soils tend to agglomerate into a mass not effectively penetrated by solvent.

Prior to extraction and drying, 50 ng $^{13}C_{12}$-2,3,7,8-TCDD and 10 ng of $^{37}Cl_4$-2,3,7,8-TCDD are added to each 10 g sample aliquot. These isotopically labeled compounds function as the internal standard (equivalent to 5.0 ppb) and the surrogate standard (equivalent to 1.0 ppb), respectively. The internal standard allows for quantitation by the isotopic dilution method. The surrogate standard permits an assessment of data quality at the 1.0 ppb level since the surrogate compound is also measured by the isotopic dilution method. Vigorous shaking or mixing for a minimum of three hours is required. The extract is filtered and concentrated prior to cleanup by column chromatography.

The first column uses silica gel, acid-modified silica gel, and base-modified silica gel to remove polar compounds. A second alumina interferences. A final polishing step based on activated charcoal provides selective fractionation of planar-type analytes ie., PCDDs, PCDFs, and non-ortho substituted PCBs).

The final measurement of TCDD in cleaned-up soil extracts utilizes capillary gas chromatography with low-resolution mass spectometry. Selected ion monitoring for masses 320, 322, and 257 were required for the native TCDD. Quantitation was based on the use of an isotopically labeled internal standard added to the sample prior to extraction. Masses 332 and 334 were monitored for the $^{13}C_{12}$-TCDD internal standard. Mass 328 was monitored for $^{37}Cl_4$-TCDD surrogate standard. Capillary columns were required which were demonstrated to resolve 2,3,7,8-TCDD from the other 21 TCDD isomers. The recommended columns were fused silica capillary columns containing cyanopropyl silicone phases such as SP-2330.

QUALITY CONTROL

Specific requirements were established to both control and assess data quality when using the previously described methodology. The requirements, described below, were specific for the Missouri Monitoring Project and are not necessarily applicable to every situation.

1. Each sample must be dosed with a known quantity of internal standard (equivalent to 2.5 ppb) and surrogate standard (equivalent to 1.0 ppb). The action limits for surrogate standard results are + 40% of the true value. Samples showing surrogate standard results outside of these limits must be reextracted and reanalyzed.

2. A laboratory "method blank" must be run along with each set of 24 or fewer samples. A method blank is performed by executing all the specified extraction and cleanup steps, except for the introduction of a 10 g sample. The method blank is also dosed with the internal standard and surrogate standard. If the blank is reported as containing 2,3,7,8-TCDD then that all samples in the set must be rerun.

3. Performance evaluation (PE) samples must be sumitted to participating labs throughout the course of the project. One type, consisting of contaminated Missouri soil or artifically dosed pottery clay, was used to assess performance prior to the contract award. Laboratories that failed to analyze these samples satisfactorily were not allowed to participate. Another type of PE sample consisted of contaminated Missouri soil that had been dried, sieved, and thoroughly homogenized. This type of sample was included in each set of 24 samples. Failure to perform satisfactorily (within + 50% of the consensus value) on this sample resulted in rejection of the entire data set.

4. Samples must be split with other participating labs on a periodic basis to assess interlaboratory variation as well as field subsampling error.

5. At least one per set of 24 samples must be run in duplicate to determine intralaboratory precision.

6. Field duplicates (individual samples taken from the same location at the same time) must be submitted periodically to determine.

6. Field duplicates (individual samples taken from the same location at the same time) must be submitted periodically to determine

7. One sample must be spiked with native 2,3,7,8-TCDD at a level of 1.0 ppb for each set of 24 or fewer samples.

8. In cases where no native 2,3,7,8-TCDD is detected, the actual detection limit must be estimated and reported based on a signal-tonoise ratio of 2.5:1 at ions 320 and 322.

9. For each sample, the internal standard must be present with at least 10:1 signal-to-noise ratio for both mass 332 and mass 334. Also, the internal standard 332:334 ratio must be within the range of 0.67-0.87.

10. "Field blanks" were provided to monitor for possible crosscontamination of samples in the field. The "field blank" consisted of uncontaminated soil (background soil taken off-site) and/or equipment rinsate (field equipment such as augers which have been rinsed with trichloroethylene or other solvent).

11. Qualitative requirements are the quality control elements that help assure that any positive data reported actually refer to 2,3,7,8-TCDD. Those requirements were as follows:

(a) Isomer specificity must be demonstrated initially and verified daily. The verification consists of injecting a mixture containing TCDD isomers that elute close to 2,3,7,8-TCDD. It contains seven TCDD isomers (2378, 1478, 1234, 1237, 1238, 1267), including those isomers known to be the most difficult to separate on SP-2330/SP-2340 columns and similar columns containing cyanoalkyl type liquid phases. The solution must be analyzed using the same chromatographic conditions and mass spectrometric conditions as were used for other samples and standards. The 2,3,7,8-TCDD must be separated from interfering isomers, with no more than a 25% valley relative to 2,3,7,8-TCDD peak.

(b) The 320/322 ratio must be within the range of 0.67 to 0.87.

(c) Ions 320, 322, and 257 must all be present and maximize together. The signal-to-mean-noise ratio must be 2.5:1 or better for all three ions.

(d) The retention time must equal (within three seconds) the retention time for the isotopically labeled 2,3,7,8-TCDD.

(e) At least one of the positive samples per set of 24 total samples must be confirmed by high-resolution mass spectrometry (resolution of 10,000 or better). Alternately, one of the positives can be confirmed by obtaining partial scan spectra from mass 150 to mass 350. The partial scan spectra guidelines are as follows:

     ° the 320/324 ratio should be 1.58 + 0.16
     ° the 257/259 ratio should be 1.03 + 0.10
     ° the 194/196 ratio should be 1.54 + 0.15

     ° ions 160, 161, 194, 196, 257, 259, 320, 322, and 324 should
       all be present with at least 5% relative abundance (relative
       to 322)

METHOD PERFORMANCE

    Based on the analyses of more than 14,000 soil samples
(including approximately 28% quality control samples), the
performance characteristics of the method have been well described[3].
The Previous statement is that method is well characterized. Giving
the preceision as "around 16% doesn't sound" well characterized.
multilaboratory analytical precision is estimated to be 16% at the
1 to 10 parts per billion level. Data for precision estimates are
summarized in Table I. The differences between duplicate sample
results for between-laboratory duplicates and blind (within-
laboratory) duplicates indicates that some sub-sampling error exists.
That is, some variability in the data may be due to incomplete mixing
of samples taken from the same site. However, the within-laboratory
duplicates did not indicate any significant sampling error for samples
taken from the same container. The primary conclusion of the duplicate
analyses QC results is that the overall uncertainty in results is
affected by sampling error as well as analytical error. Further
increases in accuracy might be achieved more readily by more thorough
mixing of samples than by improvements in the analytical method. The
performance samples are generally well-homogenized while the samples
split under field conditions are relatively inhomogenous.

    Estimates of method bias are more difficult to make because of
the lack of suitable standard reference materials containing "certified" levels of naturally incurred 2,3,7,8-TCDD. Table II summarized
estimates of bias based on the results of artifically fortified soils.
In addition, bias can be assessed through the comparison of analytical
procedures. The previously described procedure was evaluated relative
to an identical procedure involving a Soxhlet toluene extraction
procedure. No significant difference was observed for 39 soil samples
anlyzed by the two extraction techniques[3]. In a separate study,
the two procedures were evaluated in a single lab using one well-
homogenized soil sample. The Soxhlet procedure (n = 6; mean =
0.89 ppb; % RSD = 11) and the reference procedure (n = 6; mean =
0.83 ppb; % RSD = 12) again showed no significant difference.

    The frequency of "false positives" can be assessed from an
examination of data from reagent blanks and field blanks (soils known
not to contain dioxin). Only one method blank out of 201 was reported
as containing 2,3,7,8-TCDD and that level (0.14 ppb) was well below
the EPA "action" level of 1.0 ppb. Three field blanks were reported
as positve out of 131 samples at levels of 0.10, 0.01, and 0.39 ppb.

    The assessment of "false negatives" are also difficult to assess.
The qualitative controls (retention time, isotope ratios, and isomer
specificity) are considered rigorous enough to prevent this problem.
above 1.0 ppb), none have been shown to be incorrectly identified

using various confirmation techniques such as MS/MS, GC-FTIR, full scan GC/LRMS, and HRMS.

In summary, the method appears to be quite rugged, with no demonstrated bias, and has a precision of around 16% relative standard deviation in the concentrations of interest (1-100 ppb) in soil. Sampling and subsampling errors can substantially effect the precision.

SITE CLEANUP NEEDS

A number of options have been considered to remediate the more than 40 sites in Missouri[2]. Many of these options are quite expensive and require reliable and timely analytical data to minimize resource etc.) and minimize the handling of uncontaminated soil. The analytical procedure described typically requires 48-72 hours from the time of sample receipt to data reporting. This delay is considered unacceptable in many applications. Consequently, two alternate methods were developed and validated to support the cleanup needs.

The first method[6] is a modified version of the procedure previsouly described. Modifications include reduction of the sample drying time, reduction of the extraction time, extraction with 10% acetone in hexane, streamlined column chromatography, and relaxation of the isomer specificity requirement. The second method[7] is based on tandem mass spectrometry (MS/MS) and "flash" chromatography[8]. Both methods have been shown to provide data in a much more timely fashion (typically 12 hours or less for 24 samples). Method performance characteristics were found to be comparable to the conventional procedure. Comparative data are summarized in Table III.

Table 1. SUMMARY OF PRECISION FOR TCDD MEASUREMENTS IN SOIL

Quality Control Type	Number of Labs	Number of Results	Nominal Level (ppb)	% RSD
Performance Sample[a]	8	31	1	16
	8	27	4	14
	14	55	8	16
Within Lab Duplicate[b]	16	29	1 - 2	28
	16	29	2 - 5	16
	16	16	5 - 10	16
Between Lab Duplicates[c]	18	34	1 - 2	39
	18	47	2 - 5	55
	18	25	5 - 10	50

[a] Potters clay fortified with 2,3,7,8-TCDD and potential interferences (chlorinated compounds).
[b] Duplicate analyses of actual soil samples within the same laboratory. These samples are split in the lab.

c Duplicate soil samples sent to two different labs. These samples are split in the field.

Table 2. SUMMARY OF FORTIFICATION DATA

Fortification Type	Number of Labs	Number of Samples	Nominal Level (ppb)	% Bias[a]
Field Blank Spike[b]	17	378	1	-2.3
Surrogate Spike[c]	18	4030	1	+0.1

[a] Expressed as percent which reported value was higher (+) or lower (-) than the true value.
[b] Soils not containing 2,3,7,8-TCDD which were spiked before extraction by the laboratory.
[c] Added before extraction to every sample by the laboratory.

Table 3. METHOD COMPARISON STUDY

Sample[a]	Lab ID	Analytical[b] Method	Mean (ppb)	% RSD
Soil #1	A	REF	0.81	20
Soil #1	B	REF	0.75	29
Soil #1	B	REF	0.83	11
Soil #1	A	MOD	0.63	8
Soil #1	B	MOD	0.57	12
Soil #1	C	MOD	0.53	23
Soil #1	D	MS/MS	0.51	27
Soil #1	E	MS/MS	0.88	40
Soil #2	A	REF	16.6	6
Soil #2	B	REF	15.8	13
Soil #2	A	MOD	15.5	7
Soil #2	B	MOD	13.6	4
Soil #2	C	MOD	13.4	13
Soil #2	D	MS/MS	14.1	14
Soil #2	E	MS/MS	15.7	10

[a] Each sample was extracted and analyzed six times. The samples were well-homogenized, contaminated soils, from two Missouri sites.
[b] REF = Reference EPA Method; MOD = Modified LRMS Method; MS/MS = MS/MS Method.

REFERENCES

[1] Harris, D. J., Paper presented at the National Conference of Uncontrolled Hazarous Waste Sites, Washington, DC, October 1981.
[2] Kleopfer, R. D., "2,3,7,8-TCDD Contamination in Missouri," Chemosphere, Vol. 14, No. 6/7, 1985, pp 739-744.
[3] Kleopfer, R. D. and Kirchmer, C. J., "Quality Assurance Plan for 2,3,7,8-Tetrachlorodibenzo-p-dioxin Monitoring in Missouri," in Chlorinated Dioxins and Dibenzofurans in the Total Environment II, Butterworth Publishers, Stoneham, MA, 1985, pp 355-366.

[4] Kleopfer, R. D., Yue, K. T., and Bunn, W. W., "Determination of 2,3,7,8-TCDD in Soil," in Chlorinated Dioxins and Dibenzofurans in the Total Environment II, Butterworth Publishers, Stoneham, MA, 1985, pp 367-375.

[5] U.S. Environmental Protection Agency, STATEMENT OF WORK, "Dioxin Analysis; Soil/Sediment Matrix Multi-Concentration; Selected Ion Monitoring (SIM) GC/MS Analysis with Jar Extraction Procedure," September 15, 1983.

[6] Region VII, EPA, "Determination of 2,3,7,8-TCDD in Soil and Sediment Using Rapid Extraction and GC/MS," June 1985.

[7] Region VII, EPA, "Rapid Determination of TCDD in Soil and Sediment Using Gas Chromatography and Tandem Mass Spectrometry," June 1985.

[8] Smith, J. S., BenHur, D., Pavlick, R., Urban, M. J., LaFornara, J. P., Kleopfer, R. D., Yue, K. T., Kirchmer, C. J., Smith, W. A., and Viswanathan, T. S., "Comparison of a New Rapid Extraction GC/MS/MS and the Contract Laboratory Program GC/MS Methodologies for the Analysis of 2,3,7,8-TCDD," presented before the American Chemical Society, Miami, Florida, April 1985.

Forest C. Garner, Michael T. Homsher, and J. Gareth Pearson

PERFORMANCE OF USEPA METHOD FOR ANALYSIS OF 2,3,7,8-TETRACHLORO-
DIBENZO-p-DIOXIN IN SOILS AND SEDIMENTS BY CONTRACTOR LABORATORIES

REFERENCE: Garner, F. C., Homsher, M. T., and Pearson,
J. G., "Performance of USEPA Method for Analysis of 2,3,7,8-
Tetrachloro-dibenzo-p-dioxin in Soils and Sediments by Con-
tractor Laboratories," Quality Control in Remedial Site
Investigation: Hazardous and Industrial Solid Waste Testing,
Fifth Volume, ASTM STP 925, C. L. Perket, Ed., American
Society for Testing Materials, Philadelphia, 1986.

ABSTRACT: The United States Environmental Protection Agency's
superfund contractor laboratories use the method found in USEPA
IFB WA84-A002 to analyze soil and sediment samples for the specific isomer 2,3,7,8-tetrachloro-dibenzo-p-dioxin. This method
uses isotope dilution gas chromatography and mass spectrometry
(GC/MS). The precision, accuracy, detection limit, isomer specificity, recovery, and other performance parameters are important
in evaluating performance of laboratories and methods. The observed performance of this method in the production-line analysis
of thousands of samples by eleven contractor laboratories is
presented in this paper.

KEYWORDS: gas chromatography, mass spectrometry, isotope dilution, analytical method, dioxin, precision, accuracy, detection
limit.

The Environmental Protection Agency's Environmental Monitoring
Systems Laboratory - Las Vegas (EMSL-LV) is responsible for conducting
a quality assurance program in support of the Agency's Superfund Contract Laboratory Program (CLP). For additional information on the
CLP and the EMSL-LV quality assurance support program, see the papers
in this volume by Kovell[1] and Moore and Pearson[2]. The EMSL-LV

Mr. Garner is a Scientist and Mr. Homsher is a Scientific Supervisor at Lockheed Engineering and Management Services Co., P.O. Box
15027, Las Vegas, NV 89114: Mr. Pearson is the Chief of the Toxic
and Hazardous Waste Operations Branch of the USEPA Quality Assurance
Division at the Environmental Monitoring Systems Laboratory, 944 E.
Harmon, Las Vegas, NV 89109.

quality assurance support program includes providing analytical calibration standards and quality control materials, maintaining a quality
assurance data base, conducting performance evaluation studies, and
conducting audits of CLP data. Through these activities the EMSL-LV
continuously evaluates and documents CLP laboratory and method performance. Analytical methods are dynamically evaluated and validated,
i.e., the performance of the methods is continually assessed and periodically documented. Where these assessments indicate gaps in our
knowledge about the performance of the methods, the EMSL-LV designs
and conducts performance evaluation studies to fill these information
gaps. If these assessments indicate that method performance is inadequate to meet the Agency's monitoring objectives, studies are conducted
to gather the data necessary to improve method performance. The objective of this paper is to document the performance of the CLP analytical method for the analysis of 2,3,7,8-tetrachloro-dibenzo-p-dioxin
(2,3,7,8-TCDD) in soils and sediments.

BRIEF DESCRIPTION OF METHOD

The USEPA has a standard method for use by contractor laboratories
in analyzing for 2,3,7,8-TCDD in soils and sediments. This method is
described in detail in USEPA IFB WA84-A002 and by Kleopfer [3,4]. The
method is intended to achieve a detection limit of one µg/kg and to
separate 2,3,7,8-TCDD from all other TCDD isomers. Surrogate ($^{37}Cl_4$-
2,3,7,8-TCDD) and internal standard ($^{13}C_{12}$-2,3,7,8-TCDD) are added
before sample extraction. The sample is then subjected to a jar extraction technique using 150 mL hexane and 20 mL methanol in a shaker
for at least 3 hours. The extract is then subjected to cleanup using
silica gel and alumina, and, optionally, Carbopak C. The eluate is
concentrated by evaporation under a gentle stream of dry nitrogen
before GC/MS analysis. Peak areas are obtained from reconstructed
ion current profiles for m/z 257, 320, 322, 328, 332, and 334. The
quantity of native TCDD is estimated by calculating the ratio of the
sum of the peak areas for m/z 320 and 322 (native TCDD) to the sum of
the peak areas for m/z 332 and 334 (internal standard). This ratio
is then multiplied by the quantity of internal standard added to the
sample and divided by the sample weight and the mean response factor
from the 15-point initial calibration to obtain an estimate of the
concentration of native TCDD in the sample.

The following quality control analyses are routinely performed:

Routine Calibration - The lowest concentration calibration standard (corresponding to 1.0 µg/kg in a 10 gram sample) must be analyzed
once at the beginning of each 8-hour period.

Method Blank Analysis - One analysis per case (approximately 24
samples) is performed on laboratory reagents to determine if contamination is occurring.

Surrogate Accuracy - Surrogate accuracy is computed for each sample
analyzed. The surrogate percent accuracy is the quantity of surrogate

determined as a percentage of the amount added. The amount added by the contractor laboratory corresponds to 1.0 µg/kg in a 10 gram sample. The surrogate and internal standard are added simultaneously from a single standard.

Fortified Field Blank Analysis - One sample per case is fortified with 1.0 µg/kg of native TCDD by the contractor laboratory and analyzed to determine if matrix effects are present.

Duplicate Analysis - One sample per case is analyzed twice to determine within-laboratory precision.

Column Performance Check Analysis - A standard solution containing several closely eluting isomers of TCDD is analyzed initially and at the end of each 8-hour period to demonstrate adequate isomer resolution. The isomers include 2,3,7,8-TCDD, 1,2,3,4-TCDD, 1,4,7,8-TCDD, 1,2,3,7-TCDD, 1,2,3,8-TCDD, 1,2,7,8-TCDD, 1,2,6,7-TCDD.

Performance Evaluation Sample Analysis - One sample per case is a performance evaluation sample with a known amount of TCDD and potential interferences. The concentration of this sample is not known by the laboratories. Four well-characterized materials of 1 to 10 µg/kg native TCDD in dry soil are used routinely, as are two blank materials. The potentially interfering compounds are DDT, DDE, chlordane, Arochlor 1260, and 1,2,3,4-TCDD.

LIMITATIONS

The range of calibration corresponds to sample concentrations of 1.0 to 200 µg/kg TCDD. Samples exceeding 200 µg/kg must be reanalyzed using a smaller (1 g) sample aliquot to ensure accuracy, thus effectively extending the calibration range to 2000 µg/kg.

PERFORMANCE DATA

The following statistics and tables are the results of several data reductions of a comprehensive system of databases maintained at EMSL-LV. These databases include information from eleven contractor laboratories analyzing thousands of real environmental samples from hazardous waste sites across the United States under production-line conditions in commercial laboratories. The precision, accuracy, and other method performance parameters presented here may be completely inappropriate for describing method performance for analyzing meticulously prepared samples under research conditions. It should be noted that no outliers (extremely high or low values) were removed from any data sets before statistical analysis.

Precision

The precision of the method is characterized by the percent relative standard deviation (RSD), which is the standard deviation divided by the mean, and then multiplied by 100 to obtain a percentage. Precision estimates are available from fortified field blank analyses, surrogate analyses, duplicate analyses, and analyses of the four performance evaluation materials. Analysis of variance tests performed for the analyses of the four performance evaluation materials indicated that there is not significant evidence that the interlaboratory precision is different from the intralaboratory precision at the five percent level of significance. In other words, no statistical difference was observed between the laboratory means, and thus there is no significant evidence of an interlaboratory component of variance. The overall percent RSD, as presented in Table 1, was calculated using all data (irrespective of laboratory of origin) and thus may include a small amount of interlaboratory variation.

TABLE 1--Precision.

Sample Type	Number of Data	Within Laboratory %RSD	Overall %RSD
Fortified Field Blank	152	18.7	20.8
Surrogate Analysis	3520	17.2	18.1
Duplicate Analysis	33	12.5	Not Estimable
Performance Evaluation No. 1	19	13.7	13.4
Peformance Evaluation No. 2	20	12.8	16.6
Performance Evaluation No. 3	38	18.4	19.0
Performance Evaluation No. 4	43	10.6	10.2
Overall Average	3833	17.1	18.0

It should be noted that the fortified field blank samples are in various matrices, and thus exhibit somewhat high variability. The duplicate analyses represent within-laboratory replicate analyses of a single sample, and thus may include variation due to heterogeneity but cannot include interlaboratory variation. Although the concentrations of the four PE materials cannot be revealed at this time, it is apparent that the percent RSD decreases as concentration increases.

Accuracy

Method bias was estimated from fortified field blanks, surrogate analyses, and performance evaluation sample analyses. The accuracy is the mean analytical value expressed as a percentage of the true value, and the bias is the difference between the true value and the mean analytical value, expressed as a percentage of the true value. These are summarized in Table 2.

TABLE 2--Method bias.

Sample Type	Number of Data	Mean Percent Accuracy	Mean Percent Bias
Fortified Field Blank	152	96.6	-3.4
Surrogate Analysis	3520	99.5	-0.5
Performance Evaluation No. 1	19	83	-17
Performance Evaluation No. 2	20	103	+3
Performance Evaluation No. 3	38	86	-14
Performance Evaluation No. 4	43	82	-18

It is not certain which of the statistics in Table 2 are good estimates of the true accuracy and bias of the method for environmental samples. The fortified field blank and the surrogate analyses might not allow the analyte to be fully incorporated in the sample for sufficient time to properly simulate environmental TCDD. The accuracy and bias for the performance evaluation samples are based on nominal concentrations which are believed to be correct but have not been rigorously verified.

Detection Limit

Native 2,3,7,8-TCDD is detected in a sample if and only if the following criteria are met:

1. The retention time of the sample component (m/z 257, 320, and 322) must be within 3 seconds of the retention time of the internal standard.
2. The ion current for m/z 257, 320, and 322 must maximize simultaneously.
3. The integrated ion current for m/z 257, 320, 322, and 328 (surrogate) must be at least 2.5 times background noise, and must not have saturated the detector. The integrated ion current for the internal standard ions (m/z 332 and 334) must be at least 10 times background noise, and must not have saturated the detector.
4. Relative ion abundance of m/z 320 to m/z 322 must be between 0.67 and 0.87.

If native 2,3,7,8-TCDD is not detected in a sample then an estimated detection limit must be reported. The estimated detection limit is the concentration of native 2,3,7,8-TCDD corresponding to a signal-to-noise ratio of 2.5 for m/z 320 or m/z 322. This estimated detection limit may be optimistically low because:

1. The concentration corresponding to a signal-to-noise ratio of 2.5 will only average approximately 2.5, and should be expected to exceed 2.5 about half of the time, while also being less than 2.5 about half of the time. Thus if several analyses were actually performed at the estimated detection limit, only about half would achieve the required signal-to-noise criterion at m/z 320 or 322. Many definitions of detection limit specify a probability of 99% or more [5,6].
2. It only considers m/z 320 or m/z 322, while m/z 257 is usually the weakest ion (generally 20 to 45 percent of m/z 322).
3. It ignores requirements for the ratio of m/z 320 to m/z 322.

Nevertheless, very high estimated detection limits are occasionally reported. Out of 2521 reported estimated detection limits, 19 were observed between 5 and 10 µg/kg, 10 were observed between 10 and 30 µg/kg, and 1 value exceeded 30 µg/kg (202 µg/kg). The cause of these high estimated detection limits is not known, and may be due to analytical problems or very unusual sample chemistry. For this reason, values exceeding 5 µg/kg were ignored in deriving the summary statistics in Table 3.

Perhaps a better indicator of the detection limit of this method can be derived from the observed performance on known positive samples such as fortified field blanks (1 µg/kg) and performance evaluation samples (1 to 10 µg/kg). This is summarized in Table 4.

The rates of detection of the fortified field blanks and performance evaluation samples No. 1 and No. 3 may indicate that the concentration of these samples are below the method detection limit. The concentrations of the performance evaluation samples cannot be revealed in this report because these materials are still in use as blind quality control samples. However, it is apparent that the probability of detection for a 1 µg/kg sample is approximately 95 to 97 percent. A probability of detection of 99 percent is probably achieved at approximately 5 to 10 µg/kg. It should be pointed out that the

TABLE 3--Estimated detection limits.

	Sample Mean	Standard Deviation	Upper Percentiles*			
			50%	90%	95%	99%
Estimated Detection Limit (µg/kg)	0.30	0.53	0.11	0.71	1.0	3.4

Based on values between zero and five µg/kg (N = 2491).
*For example, 90% of the reported estimated detection limits were less than or equal to 0.71 µg/kg, while 99% were less than or equal to 3.4 µg/kg. If the values greater than 5 µg/kg were included, then each percentile would occur at a slightly higher estimated detection limit.

TABLE 4--Detection of known positive samples.

Sample Type	Number of Analyses	Number Detected	Probability of Detection
Fortified Field Blanks	152	147	97%
Performance Evaluation No. 1	20	19	95%
Performance Evaluation No. 2	20	20	100%
Performance Evaluation No. 3	41	39	95%
Performance Evaluation No. 4	43	43	100%

false negative results on the performance evaluation samples all came from the same laboratory, and thus may reflect a laboratory problem rather than a method problem.

Sample Cleanup Experiment Results

Three sample cleanup columns are used: silica gel, alumina, and Carbopak C. The use of these columns reduces the amount of TCDD (native, surrogate, and internal standard) injected into the GC/MS. A special experiment was performed to determine the recovery of TCDD after each cleanup column. This is completely independent of extraction recovery, as the experiment was performed using a calibration standard solution and no extraction step was involved. The percent recovery was determined by using $^{13}C_{12}$-1,2,3,4-TCDD as a recovery standard added just before injection. These are summarized in Table 5.

TABLE 5--Cleanup recoveries (mean ± standard deviation).

	Silica Cleanup	Alumina Cleanup	Carbopak C Cleanup	Silica and Alumina	Silica, Alumina, and Carbon
Percent Recovery	89.3±14.8	86.0±13.2	84.6±19.8	78.5±16.3	62.7±24.1

Each of the recoveries in Table 5 is a mean of 24 analytical results (three analyses on each column combination by each of eight laboratories). The mean percent recovery with no sample cleanup was
101.1± 9.3, based on 24 analyses. Note the large standard deviation
associated with the Carbopak C cleanup.

This experiment also permitted the investigation of the correlations between the TCDD value, the surrogate percent accuracy, and the
percent recovery.

TABLE 6--Linear correlation coefficients.

	TCDD Value	Surrogate % Accuracy	Percent Recovery
TCDD Value	1.00	0.329	-0.163
Surrogate % Accuracy	0.329	1.00	-0.286
Percent Recovery	-0.163	-0.286	1.00

Table 6 presents the means of the linear (Pearson) correlation
coefficients for each laboratory. These means (0.329, -0.163, and
-0.286) are not statistically significantly different from zero, but
may nevertheless indicate weak relationships. Note that the percent
recovery is very poorly correlated with the TCDD value, which reflects
the statistical independence which is characteristic of isotope dilution methods. Also note the weak correlation of TCDD value with the
surrogate percent accuracy. This raises substantial doubt that the
surrogate percent accuracy can be used as a reliable indicator of the
accuracy of native TCDD analysis.

Isomer Resolution

The column performance check analysis is designed to demonstrate
adequate isomer specificity between TCDD isomers. The resolution
is measured by the percent valley. The percent valley is the ion current intensity (height of baseline) between 2,3,7,8-TCDD and the next
closest peak, expressed as a percentage of the ion current intensity

(peak height) of 2,3,7,8-TCDD. The IFB WA84-A002 specifies that the percent valley must be less than 25% to demonstrate adequate isomer specificity.

The actual percent valley observed depends greatly on which GC column is used. The distribution of the percent valley statistic is positively skewed to the extent that interpretation of the mean and standard deviation are difficult. For this reason, the median (50th percentile) and upper 90th percentile are more appropriate descriptors than the mean and standard deviation. For example, as one might observe from Table 7, half of the 572 percent valleys from the CP SIL 88 column were less than or equal to 8%, while 90% of them were less than or equal to 24%. It may also be possible to resolve TCDD isomers using other columns, such as a 60 meter DB-5 [7].

TABLE 7--Percent valley by GC column.

Column Used	Number of Data	Median Percent Valley	Upper 90th Percentile
CP SIL 88*	572	8	24
SP 2330*	570	12	28
SP 2340	171	12	30

*Recommended for use in IFB WA84-A002

False Positives

Reagent blank analysis results may be used to estimate the rate of false positives in "clean" samples. Out of 210 reagent blank analyses, 2 were observed containing native 2,3,7,8-TCDD. Thus one might speculate that approximately 1 percent of "clean" environmental samples may be falsely reported to contain native TCDD at detectable levels. The source of these positives is not known, but they are probably not due to incomplete labelling of the surrogate or internal standard. Incomplete labelling would result in a mass ratio (m/z 320 to 322) substantially different from that required to identify native TCDD.

It is noteworthy that both observed false positives occurred at extremely low levels (0.024 and 0.026 µg/kg).

Conclusions

The precision of the method is characterized by a relative standard deviation of approximately 10 to 20 percent for concentrations of 1.0 to 10 µg/kg. It is not clear if the method is biased - the various sample types indicate accuracies varying from approximately 80 to 100 percent of the true value. The estimated detection limit has a mean of approximately 0.30 µg/kg and a median of approximately 0.11 µg/kg, while the probability of detection of a 1.0 µg/kg sample is approximately 95 to 97 percent. The two GC columns recommended in IFB WA84-A002 are capable of meeting the 25 percent valley criterion approximately 90 percent of the time.

ACKNOWLEDGMENTS

Although the research described in this article has been funded wholly or in part by the United States Environmental Protection Agency through contract number 68-03-3249 to Lockheed Engineering and Management Services Company, it has not been subjected to Agency review and therefore does not necessarily reflect the views of the Agency and no official endorsement should be inferred. Mention of trade names or commercial products does not constitute endorsement or recommendation for use.

The authors are grateful for the invaluable assistance provided by Mary Flynn, Jordan Fan, John Fountain, and Shirley Schieck in gathering and analyzing contractor laboratory data. The authors are also indebted to Yves Tondeur, G. Wayne Sovocool, and L. R. Williams for technical and editorial reviews of this paper.

REFERENCES

[1] Kovell, S. P., "Contract Laboratory Program - An Overview," Quality Control in Remedial Site Investigation: Hazardous and Industrial Solid Waste Testing, Fifth Volume, ASTM STP 925, C. L. Perket, Ed., American Society for Testing Materials, Philadelphia, 1986.
[2] Moore, J. M., and Pearson, J. G., "Quality Assurance Support for the Superfund Contract Laboratory Program," Quality Control in Remedial Site Investigation: Hazardous and Industrial Solid Waste Testing, Fifth Volume, ASTM STP 925, C. L. Perket, Ed., American Society for Testing Materials, Philadelphia, 1986.
[3] Invitation for Bid, Solicitation no. WA84-A002, United States Environmental Protection Agency, 400 M Street SW, Washington, D.C., 20460, 1984.
[4] Kleopfer, R., "Dioxin Analytical Methods - General Description and Quality Control Considerations," Quality Control in Remedial Site Investigation: Hazardous and Industrial Solid Waste Testing, Fifth Volume, ASTM STP 925, C. L. Perket, Ed., American Society for Testing Materials, Philadelphia, 1986.
[5] Long, G. L., and Winefordner, J. D., "Limit of Detection: A Closer Look at the IUPAC Definition," Analytical Chemistry, Vol. 55, No. 7, June 1983, pp. 712A-724A.

[6] Keith, L. H., Crummett, W., Deegan, J., Libby, R. A., Taylor, J. K., and Wentler, G., "Principles of Environmental Analysis," Analytical Chemistry, Vol. 55, No. 14, December 1983, pp. 2210-2218.

[7] Solch, J. G., et al., "Analytical Methodology for Determination of 2,3,7,8-Tetrachlorodibenzo-p-dioxin in Soils," Chlorinated Dixoins and Dibenzofurans in the Total Environment," Vol. II, L. H. Keith, C. Rappe, and G. Choudhary, Eds., Butterworth Publishers, Boston, 1985.

Joan F. Fisk

SEMI-VOLATILE ORGANIC ANALYTICAL METHODS - GENERAL
DESCRIPTION AND QUALITY CONTROL CONSIDERATIONS.

REFERENCE: Fisk, J. F., "Semi-Volatile Organic Analytical
Methods — General Description and Quality Control Consider-
ations," Quality Control in Remedial Site Investigation:
Hazardous and Industrial Solid Waste Testing, Fifth Volume,
ASTM STP 925, C.L. Perket, Ed., American Society for
Testing and Materials, 1986.

ABSTRACT: As a result of the enactment of the Comprehensive Environmental Response, Compensation and Liability
Act (CERCLA) in 1980, and the subsequent delegation of its
authority by the President of the United States to the
Administrator of the Environmental Protection Agency, the
Contract Laboratory Progam (CLP) has been developed to
analyze "Superfund" samples in a broad-based manner in
order to obtain the most information about them with a
reasonable investment in time and money. The CLP uses gas
chromatography/mass spectrometry (GC/MS) as its primary
tool for analysis of samples for organic constituents. More
explicitly, it employs solvent extraction techniques respectively appropriate for water and soil/sediment matrices prior
to MS detection for semi-volatile compounds. An elaborate
Quality Assurance/Quality Control (QA/QC) system is in
place to guarantee that data generated by the CLP is of
known quality and will hold up to the rigors of litigation, in
addition to providing an ongoing monitoring of CLP laboratories QA/QC data in order to guarantee their successful
performance.

KEYWORDS: CERCLA, Contract Laboratory Program (CLP),
Superfund, Gas Chromatography/Mass Spectrometry (GC/
MS), extraction, Quality Assurance/Quality Control (QA/QC),
surrogates, matrix spikes, internal standards, System Performance Check Compounds (SPCC), Calibration Check Compounds (CCC).

Joan Fisk is a Project Officer/Chemist with the Environmenal Protection
Agency, Office of Solid Waste and Emergency Response, Office of
Emergency and Remedial Response, Hazardous Response Support Division, Analytical Support Branch, 401 M Street, S.W., Washington, D.C.,
20460

INTRODUCTION

Public Law 96-510 entitled the Comprehensive Environmental Response, Compensation and Liability Act (CERCLA) was enacted in 1980 and bestowed upon the President of the United States the authority to effect the removal from the environment of any hazardous substances, pollutant, or contaminant in order to protect public health and welfare and the environment. The President delegated his authority to the Administrator of the Environmental Protection Agency (EPA) upon passage of CERCLA - or better known as "Superfund" - and thus began the saga of the Contract Laboratory Program, or CLP.

In order to quickly identify analytes in samples which are of a hazardous or polluting nature it is necessary to have a system in place to quickly analyze samples generated from the National Priority List (NPL) sites (or Superfund sites) by routine standardized methods and to produce the data for these samples in a uniform manner which can clearly identify the compounds of concern.

Another major facet of CLP analysis is the need for a rigid Quality Assurance/Quality Control (QA/QC) program built into CLP procedures. EPA's Environmental Monitoring Systems Laboratory in Las Vegas (EMSL-LV) monitors the QA/QC of the CLP and maintains an extraordinary QA/QC data base which enables them to identify problems within both CLP methods and CLP laboratories. In addition, the QA/QC requirements that are an integral part of the protocols put the stamp of authenticity on all data generated by CLP labs so that data is of known quality if and when it reaches a court of law and can serve as a witness to liability for the violation of our environment.

The purpose of this paper is to describe the methods used to analyze Superfund samples for semi-volatile compounds, the reasons for using a broad-based approach, the Quality Assurance/Quality Control requirements contained in the methods, and the results of this unique approach of the CLP.

The specific analyses described herein are for semi-volatile organics analysis using selective extraction procedures prior to gas chromotography (GC) for the separation of analytes and mass spectrometry (MS) for detection of those analytes. For detailed protocols, the reader should see the most recent version of CLP methods for organics analysis (1).

A brief historical profile follows, as well as a fairly explicit description of the protocols used for analyses of semi-volatile compounds. The importance of the procedural QA/QC requirements in generating data of known quality by all CLP labs will be discussed.

BACKGROUND

Until October 1984, CLP methods for semi-volatile analyses were based on EPA method 625, (2)(3)(4) with revisions over the years to make them more suitable for samples of varying and often extremely difficult matrices. In October 1984, the CLP modified all contracts to utilize protocols called the COP (Consensus Organic Protocols) which emerged through a series of caucuses with technical representation from EPA Regions, CLP Contractors, EMSL-LV and EMSL-Cincinnati, and chaired by EPA Headquarters Project Officers. In addition to the caucus participants, the laboratory community across the nation has requested these methods.

SCOPE OF ANALYSIS

Since CERCLA defines a hazardous substance as any substance which may present substantial danger to public health, welfare, or the environment and includes substances referenced in the Water Pollution Control Act, the Toxic Substances Control Act, the Clean Air Act, and the Solid Waste Disposal Act, the number of possible compounds for identification is almost infinite in terms of real-time analysis. In order to get around a potentially forbidding task, the CLP has identified a list of compounds in Exhibit C of CLP Statements of Work which it has designated as the Hazardous Substance List (HSL). This is a misnomer and should be called the Target Compound List meaning the compounds which are routinely searched for during organics analysis. This list is dynamic and can be altered to include new compounds of concern or to delete compounds for which the protocols are deemed to be inappropriate (as determined by performance based QA/QC). Following in Table 1 is the list of Semi-Volatile Target Compounds and their associated Contract Required Detection Limits (CRDLs).

To broaden the scope of the GC/MS analysis for semi-volatiles, the methods require a search of the EPA/NIH National Standard Reference Data System (commonly called the NBS library) to tentatively identify the spectrum of up to 20 of the largest GC peaks in the reconstructed ion current (RIC) chromatograms for each sample.

SCREENING OF SEMI-VOLATILE SAMPLES

It is required to characterize CLP soil/sediment samples prior to GC/MS analysis in order to select the appropriate sample preparation protocol (and to prevent instrument downtime due to swamping the GC/MS. It is recommended to screen water samples in order to prevent instrument downtime.

Following is a summary of recommended screening procedures which may be used to characterize the sample prior to choosing the appropriate analytical method (soil/sediment) or the right dilution (aqueous).

Sample preparation methods for screening are described under the "Semi-Volatiles Organic Compounds Analysis" section (following). The extracts produced from the three sample preparations described (aqueous, low level soil and medium level soil) are screened by GC/FID using an external GC calibration standard (phenol, phenanthrene and di-n-octyl phthalate). Designated aliquots of the extracts for each of the three categories are injected for screening.

TABLE 1 -- Semi-Volatile Target Compounds

Semi-Volatiles	CAS Number	Contract Required Detection Limits	
		Low Water[a] ug/L	Low Soil/Sediment[b] ug/Kg
1. Phenol	108-95-2	10	330
2. bis(2-Chloroethyl) ether	111-44-4	10	330
3. 2-Chlorophenol	95-57-8	10	330
4. 1,3-Dichlorobenzene	541-73-1	10	330
5. 1,4-Dichlorobenzene	106-46-7	10	330
6. Benzyl Alcohol	100-51-6	10	330
7. 1,2-Dichlorobenzene	95-50-1	10	330
8. 2-Methylphenol	95-48-7	10	330
9. bis(2-Chloroisopropyl) ether	39638-32-9	10	330
10. 4-Methylphenol	106-44-5	10	330
11. N-Nitroso-Dipropylamine	621-64-7	10	330
12. Hexachloroethane	67-72-1	10	330
13. Nitrobenzene	98-95-3	10	330
14. Isophorone	78-59-1	10	330
15. 2-Nitrophenol	88-75-5	10	330
16. 2,4-Dimethylphenol	105-67-9	10	330
17. Benzoic Acid	65-85-0	50	1600
18. bis(2-Chloroethoxy) methane	111-91-1	10	330
19. 2,4-Dichlorophenol	120-83-2	10	330
20. 1,2,4-Trichlorobenzene	120-82-1	10	330
21. Naphthalene	91-20-3	10	330
22. 4-Chloroaniline	106-47-8	10	330
23. Hexachlorobutadiene	87-68-3	10	330
24. 4-Chloro-3-methylphenol (para-chloro-meta-cresol)	59-50-7	10	330
25. 2-Methylnaphthalene	91-57-6	10	330
26. Hexachlorocylopentadiene	77-47-4	10	330
27. 2,4,6-Trichlorophenol	88-06-2	10	330
28. 2,4,5-Trichlorophenol	95-95-4	50	1600
29. 2-Chloronaphthalene	91-58-7	10	330
30. 2-Nitroaniline	88-74-4	50	1600
31. Dimethyl Phthalate	131-11-3	10	330
32. Acenaphthylene	208-96-8	10	330
33. 3-Nitroaniline	99-09-2	50	1600
34. Acenaphthene	83-32-9	10	330
35. 2,4-Dinitrophenol	51-28-5	50	1600

TABLE 1 -- Semi-Volatile Target Compounds (continued)

Semi-Volatiles	CAS Number	Contract Required Detection Limits	
		Low Water[a] ug/L	Low Soil/ Sediment[b] ug/Kg
36. 4-Nitrophenol	100-02-7	50	1600
37. Dibenzofuran	132-64-9	10	330
38. 2,4-Dinitrotoluene	121-14-2	10	330
39. 2,6-Dinitrotoluene	606-20-2	10	330
40. Diethylphthalate	84-66-2	10	330
41. 4-Chlorophenyl Phenyl ether	7005-72-3	10	330
42. Fluorene	86-73-7	10	330
43. 4-Nitroaniline	100-01-6	50	1600
44. 4,6-Dinitro-2-methylphenol	534-52-1	50	1600
45. N-nitrosodiphenylamine	86-30-6	10	330
46. 4-Bromophenyl Phenyl ether	101-55-3	10	330
47. Hexachlorobenzene	118-74-1	10	330
48. Pentachlorophenol	87-86-5	50	1600
49. Phenanthrene	85-01-8	10	330
50. Anthracene	120-12-7	10	330
51. Di-n-butylphthalate	84-74-2	10	330
52. Fluoranthene	206-44-0	10	330
53. Pyrene	129-00-0	10	330
54. Butyl Benzyl Phthalate	85-68-7	10	330
55. 3,3'-Dichlorobenzidine	91-94-1	20	660
56. Benzo(a)anthracene	56-55-3	10	330
57. Bis(2-ethylhexyl)phthalate	117-81-7	10	330
58. Chrysene	218-01-9	10	330
59. Di-n-octyl Phthalate	117-84-0	10	330
60. Benzo(b)fluoranthene	205-99-2	10	330
61. Benzo(k)fluoranthene	207-08-9	10	330
62. Benzo(a)pyrene	50-32-8	10	330
63. Indeno(1,2,3-cd)pyrene	193-39-5	10	330
64. Dibenz(a,h)anthracene	53-70-3	10	330
65. Benzo(g,h,i)perylene	191-24-2	10	330

[a] Medium Water Contract Required Detection Limits (CRDL) for Semi-Volatile HSL Compounds are 100 times the individual Low Water CRDL.

[b] Medium Soil/Sediment Contract Required Detection Limits (CRDL) for Semi-Volatile HSL Compounds are 60 times the individual Low Soil/Sediment CRDL.

TABLE 1 -- Semi-Volatile Target Compounds (continued)

Pesticides[a]	CAS Number	Contract Required Detection Limits	
		Low Water[b] ug/L	Low Soil/ Sediment[c] ug/Kg
66. alpha-BHC	319-84-6	0.05	8.0
67. beta-BHC	319-85-7	0.05	8.0
68. delta-BHC	319-86-8	0.05	8.0
69. gamma-BHC (Lindane)	58-89-9	0.05	8.0
70. Heptachlor	76-44-8	0.05	8.0
71. Aldrin	309-00-2	0.05	8.0
72. Heptachlor Epoxide	1024-57-3	0.05	8.0
73. Endosulfan I	959-98-8	0.05	8.0
74. Dieldrin	60-57-1	0.10	16.0
75. 4,4'-DDE	72-55-9	0.10	16.0
76. Endrin	72-20-8	0.10	16.0
77. Endosulfan II	33213-65-9	0.10	16.0
78. 4,4'-DDD	72-54-8	0.10	16.0
79. Endosulfan Sulfate	1031-07-8	0.10	16.0
80. 4,4'-DDT	50-29-3	0.10	16.0
81. Endrin Ketone	53494-70-5	0.10	16.0
82. Methoxychlor	72-43-5	0.5	80.0
83. Chlordane	57-74-9	0.5	80.0
84. Toxaphene	8001-35-2	1.0	160.0
85. AROCLOR-1016	12674-11-2	0.5	80.0
86. AROCLOR-1221	11104-28-2	0.5	80.0
87. AROCLOR-1232	11141-16-5	0.5	80.0
88. AROCLOR-1242	53469-21-9	0.5	80.0
89. AROCLOR-1248	12672-29-6	0.5	80.0
90. AROCLOR-1254	11097-69-1	1.0	160.0
91. AROCLOR-1260	11096-82-5	1.0	160.0

[a] Pesticides are target compounds in the semi-volatile fraction when at a concentration high enough to be detected by GC/MS.

[b] Medium Water Contract Required Detection Limits (CRDL) for Pesticide HSL Compounds are 100 times the individual Low Water CRDL.

[c] Medium Soil/Sediment Contract Required Detection Limits (CRDL) for Pesticide HSL Compounds are 15 times the individual Low Soil/Sediment CRDL.

Water Samples

o   If no sample peaks are detected or all are less than 100% full scale deflection, the extract prepared for the water analysis is analyzed by GC/MS undiluted.

o   If any sample peaks are greater than 100% full scale deflection, the appropriate dilution to reduce peaks to between 50-100% full scale deflection is made prior to GC/MS analyses.

Soil/Sediment Samples

o   If no sample peaks are detected from either low or medium extract (for the low level screen use 5 ml of the 300 ml total extract from the 30 g sample extraction prior to concentration, and for the medium level screen use 5 ml from the total 10 ml extract of the 1 g sample concentrated to 1 ml) or all are less than 10% full peak deflection, the low level sample preparation is the appropriate one.

o   If peaks are detected at greater than 10% deflection and less than or equal to 100% full scale deflection, the medium level extract is the appropriate one to analyze with any necessary dilution.

o   If peaks are detected at greater than 100% full scale deflection, the medium level preparation with dilution to bring peaks to between 50-100% full scale deflection is the correct extract to use.

SEMI-VOLATILE ORGANIC COMPOUNDS ANALYSIS (5)

Sample Preparation. Three sample preparation procedures are used to allow for differing sample matrices and the concentration level anticipated. These procedures are summarized as follows:

(1)   Water samples - Approximately one liter of water is serially extracted by separatory funnel three times with methylene chloride (or alternately by continuous liquid-liquid extraction) at pH greater than 11 first, and then at pH less than 2. The extracts are concentrated separately by Kuderna-Danish evaporation and are combined in equal portions just prior to injection into the GC/MS.

(2)   Medium level soil/sediment samples - Two grams of anhydrous sodium sulfate are mixed with a one gram soil sample from which water has been decanted and discarded, and the mixture is extracted with methylene chloride by disruption with an ultrasonic probe. The extract is filtered and then concentrated by nitrogen blowdown prior to GC/MS analysis.

(3)   Low level soil/sediment samples - 30 grams of sample are mixed with 60 grams of anhydrous sodium sulfate and extracted with methylene chloride/acetone using an ultrasonic probe. The extract is filtered and concentrated by Kuderna-Danish evaporation prior to GC/MS analysis.

Gas Chromatography and Mass Spectrometric Detection. A 30 m x 0.25-0.32 mm i.d. fused silica capillary column (FSCC) is used for chromatographic analysis. For detection, the mass spectrometer is scanned from 35 to 500 amu every second or less at 70 eV in the EI ionization mode. The mass spectrum for decafluorotriphenylphosphine (DFTPP) must meet the criteria in Table 2 prior to, or simultaneous with, the analysis of a standard, and prior to analysis of extracts of blanks or samples.

TABLE 2 -- DFTPP Key Ions and Ion Abundance Criteria

Mass	Ion Abundance Criteria
51	30.0 to 60.0 percent of mass 198
68	less than 2.0 percent of mass 69
70	less than 2.0 percent of mass 69
127	40.0 - 60.0 percent of mass 198
197	less than 1.0 percent of mass 198
198	base peak, 100 percent relative abundance
199	5.0 - 9.0 percent of mass 198
275	10.0 - 30.0 percent of mass 198
365	greater than 1.00 percent of mass 198
441	present but less than mass 443
442	greater than 40.0 percent of mass 198
443	17.0 - 23.0 percent of mass 442

Qualitative Identification. There are three requirements for identification of a semi-volatile compound (in water, low level soil/sediment, or medium level soil/sediment samples). These requirements are as follows:

(1) All m/z's present in the standard mass spectrum at a relative intensity greater than 10% of the most abundant m/z in the spectrum (100%) must be present in the sample spectrum.

(2) The relative intensities of the m/z's that are greater than 10% of the most abundant m/z must agree within +/- 20% between the standard and sample spectrum.

(3) M/z's greater than 10% relative intensity in the sample spectrum but not in the standard spectrum must be considered and accounted for by the mass spectral interpretor.

Quantification. Semi-volatile target compounds are quantified using multiple internal standards, with the internal standard to be used designated as the one nearest in retention time to that of a given analyte, as given in Table 3. The extracted ion current profile (EICP) area at the characteristic m/z of a given target compound analyte is used to calculate its concentration. The characteristic m/z's are given in Tables 4, 5 and 6.

TABLE 3 -- Semi-Volatile Internal Standards with Corresponding Analytes Assigned for Quantitation

1,4-Dichlorobenzene-d_4	Naphthalene-d_8	Acenaphthene-d_{10}	Phenanthrene-d^{10}	Chrysene-d_{12}	Perylene-d_{12}
Phenol	Nitrobenzene	Hexachlorocyclopentadiene	4,6-Dinitro-2-methylphenol	Pyrene	Di-n-octyl Phthalate
bis(2-Chloroethyl) ether	Isophorone	2,4,6-Trichlorophenol	N-nitrosodiphenylamine	Butylbenzyl Phthalate	Benzo(b)fluoranthene
2-Chlorophenol	2-nitrophenol	2,4,5-Trichlorophenol	1,2-Diphenylhydrazine	3,3'-Dichlorobenzidine	Benzo(k)fluoranthene
1,3-Dichlorobenzene	2,4-Dimethylphenol	2-Chloronaphthalene	4-Bromophenyl Phenyl Ether	Benzo(a)anthracene	Benzo(a)pyrene
1,4-Dichlorobenzene	Benzoic acid	2-Nitroaniline	Hexachlorobenzene	bis(2-ethylhexyl) Phthalate	Indeno(1,2,3-cd) pyrene
Benzyl Alcohol	bis(2-Chloroethoxy)methane	Dimethyl Phthalate	Pentachlorophenol	Chrysene	Dibenz(a,h) anthracene
1,2-Dichlorobenzene	2,4-Dichlorophenol	Acenaphthylene	Phenanthrene	Terphenyl-d_{14} (surr)	Benzo (g,h,i) perylene
2-Methylphenol	1,2,4-Trichlorobenzene	3-Nitroaniline	Anthracene		
bis(2-Chloroisopropyl)ether	Naphthalene	Acenaphthene	Di-n-butyl Phthalate		
4-Methylphenol	4-Chloroaniline	2,4-Dinitrophenol	Fluoranthene		
N-nitroso-Di-n-propylamine	Hexachlorobutadiene	4-Nitrophenol			
Hexachloroethane	4-Chloro-3-methylphenol	Dibenzufuran			
2-Fluorophenol (surr)	2-Methylnaphthalene	2,4-Dinitrotoluene			
Phenol-d_6 (surr)	Nitrobenzene-d_5 (surr)	2,6-Dinitrotoluene			
		Diethyl Phthalate			
		4-Chlorophenylphenyl ether			
		Fluorene			
		4-Nitroaniline			
		2-Fluorobiphenyl (surr)			
		2,4,6-Tribromo Phenol (surr)			

surr = Surrogate Compound

TABLE 4 -- Characteristic Ions for Surrogates and
Internal Standards for Semi-Volatile Organic Compounds

Compound	Primary Ion	Secondary Ion(s)
SURROGATES		
Phenol-d_5	99	42, 71
2-Fluorophenol	112	64
2,4,6-Tribromophenol	330	332, 141
d-5 Nitrobenzene	82	128, 54
2-Fluorobiphenyl	172	171
Terphenyl	244	122, 212
INTERNAL STANDARDS		
1,4-Dichlorobenzene-d_4	152	115
Naphthalene-d_8	136	68
Acenapthene-d_{10}	164	162, 160
Phenanthrene-d_{10}	188	94, 80
Chrysene-d_{12}	240	120, 236
Perylene-d_{12}	264	260, 265

TABLE 5 -- Characteristic Ions for Pesticides/PCBs

Parameter	Primary Ion	Secondary Ion(s)
Alpha-BHC	183	181, 109
Beta-BHC	181	183, 109
Delta-BHC	183	181, 109
Gamma-BHC (Lindane)	183	181, 109
Heptachlor	100	272, 274
Aldrin	66	263, 220
Heptachlor Epoxide	353	355, 351
Endosulfan I	195	339, 341
Dieldrin	79	263, 279
4,4'-DDE	246	248, 176
Endrin	263	82, 81
Endosulfan II	337	339, 341
4,4'-DDD	235	237, 165
Endosulfan Sulfate	272	387, 422
4,4'-DDT	235	237, 165
Methoxychlor	227	228
Chlordane	373	375, 377
Toxaphene	159	231, 233
Aroclor-1016	222	260, 292
Aroclor-1221	190	222, 260
Aroclor-1232	190	222, 260
Aroclor-1242	222	256, 292
Aroclor-1248	292	362, 326
Aroclor-1254	292	362, 326
Aroclor-1260	360	362, 394
Endrin Ketone	317	67, 319

TABLE 6 -- Characteristic Ions for Semi-Volatile Target Compounds

Parameter	Primary Ion	Secondary Ion(s)
Phenol	94	65, 66
bis(-2-Chloroethyl) Ether	93	63, 95
2-Chlorophenol	128	64, 130
1,3-Dichlorobenzene	146	148, 113
1,4-Dichlorobenzene	146	148, 113
Benzyl Alcohol	108	79, 77
1,2-Dichlorobenzene	146	148, 113
2-Methylphenol	108	107
bis(2-chloroisopropyl) Ether	45	77, 79
4-Methylphenol	108	107
N-Nitroso-Di-Propylamine	70	42, 101, 130
Hexachloroethane	117	201, 199
Nitrobenzene	77	123, 65
Isophorone	82	95, 138
2-Nitrophenol	139	65, 109
2,4-Dimethylphenol	122	107, 121
Benzoic Acid	122	105, 77
bis(2-Chloroethoxy)Methane	93	95, 123
2,4-Dichlorophenol	162	164, 98
1,2,4-Trichlorobenzene	180	182, 145
Naphthalene	128	129, 127
4-Chloroaniline	127	129
Hexachlorobutadiene	225	223, 227
4-Chloro-3-Methylphenol	107	144, 142
2-Methylnaphthalene	142	141
Hexachlorocyclopentadiene	237	235, 272
2,4,6-Trichlorophenol	196	198, 200
2,4,5-Trichlorophenol	196	198, 200
2-Chloronaphthalene	162	164, 127
2-Nitroaniline	65	92, 138
Demethyl Phthalate	163	194, 164
Acenaphthylene	152	151, 153
3-Nitroaniline	138	108, 92
Acenaphthene	153	152, 154
2,4-Dinitrophenol	184	63, 154
4-Nitrophenol	139	109, 65
Dibenzofuran	168	139
2,4-Dinitrotoluene	89	63, 182
2,6-Dinitrotoluene	165	89, 121
Diethylphthalate	149	177, 150
4-Chlorophenyl-phenylether	204	206, 141
Fluorene	166	165, 167
4-Nitroaniline	138	92, 108
4,6-Dinitro-2-Methylphenol	198	182, 77
N-Nitrosodiphenylamine	169	168, 167
4-Bromophenyl-phenylether	248	250, 141
Hexachlorobenzene	284	142, 249
Pentachlorophenol	266	264, 268
Phenanthrene	178	179, 176
Anthracene	178	179, 176
Di-N-Butylphthalate	149	150, 104

TABLE 6 -- Characteristic Ions for Semi-Volatile Target Compounds (continued)

Parameter	Primary Ion	Secondary Ion(s)
Fluoranthene	202	101, 100
Pyrene	202	101, 100
Butylbenzylphthalate	149	91, 206
3,3'-Dichlorobenzidine	252	254, 126
Benzo(a)Anthracene	228	229, 226
bis(2-Ethylhexyl)Phthalate	149	167, 279
Chrysene	228	226, 229
Di-N-Octyl Phthalate	149	—
Benzo(b)Fluoranthene	252	253, 125
Benzo(k)Fluoranthene	252	253, 125
Benzo(a)Pyrene	252	253, 125
Indeno(1,2,3-cd)Pyrene	276	138, 227
Dibenz(a, h)Anthracene	278	139, 279
Benzo(g, h, i)Perylene	276	138, 277

The equations for determining the concentration of an analyte are as follows: Note that pesticide quantification by GC/MS is required when concentrations found by GC/EC are great enough to warrant MS information in the B/N fraction.

(1) <u>For low and medium level water samples:</u>

$$\text{Concentration (ug/L)} = (A_x)(I_s)(V_t)/(A_{is})(RF)(V_o)(V_i)$$

where: A_x = EICP area at the characteristic m/z of the analyte

A_{is} = EICP area at the characteristic m/z of the specified internal standard

I_s = amount of internal standard injected (ng)

V_o = volume of water extracted (mL)

V_i = volume of extract injected (uL)

V_t = volume of total extract (2,000 uL or a fraction of this volume when dilutions are made. The 2,000 uL is derived from combining half of the 1 mL base/neutral extract and half of the 1 mL acid extract)

(2) <u>For Soil/Sediment Samples:</u>

$$\text{Concentration (ug/kg)} = (A_x)(I_s)(V_t)/(A_{is})(RF)(V_i)(W_s)(D)$$

where: A_x, I_s, A_{is} are as above, and

V_t = volume of low level extract (1,000 uL or appropriate fraction) or medium level extract (2,000 uL or appropriate fraction)

V_i = volume of extract injected (uL)

D = 100 - % moisture (calculated on dry wt. basis)

W_s = weight of sample extracted (g)

The response factor (RF) is obtained from the appropriate daily standard analysis and is calculated from the equation:

$$RF = (A_x)(C_{is})/(A_{is})(C_x)$$

where: A_x = EICP area at the characteristic m/z of the analyte

A_{is} = EICP area at the characteristic m/z of the specified internal standard

C_{is} = concentration of the internal standard

C_x = concentration of the analyte in the standard

QUALITY ASSURANCE/QUALITY CONTROL

As mentioned in the introduction, the level of QA/QC attached to the CLP methods assures that the data produced are of a known quality and can be related to EPA Data Quality Objectives (DQOs) which are qualitative and quantitative statements developed by data users to define the quality of the data needed for a particular data collection activity to support specific remedial decisions or regulatory actions. The QA/QC aspects may be divided into two categories, namely, instrument or system performance QA/QC and method performance or sample QA/QC. Note that the QA/QC criteria described in the methods are contract requirements and are not to be construed as technical requirements.

Instrument QA/QC

o   Prior to analysis of any blanks or samples and simultaneous with or prior to analysis of standards, the ion criteria for decafluoro-triphenylphosphine (DFTPP) must be met as defined earlier in Table 2. The DFTPP tune must be successfully demonstrated every 12 hours of analysis time.

o   Five-point initial calibrations at designated concentrations are required prior to any analysis of blanks or samples to define the dynamic range of the GC/MS system. Certain compounds have been selected as System Performance Check Compounds (SPCC) and must have a minimum average response factor ($\overline{RF}$) of 0.05. Other compounds have been designated as Calibration Check Compounds (CCC) and must have a percent relative standard deviation (% RSD) of less than 30% to ensure the validity of the initial calibration. Only when these SPCC and CCC requirements are met is the instrument calibration considered valid.

o   After 12 hours of analysis, a continuing calibration (at a designated concentration) must be performed following a successful DFTPP tune. Again, the minimum RF allowed is 0.05 for SPCCs. For CCCs the RF must not deviate more than 25% from the $\overline{RF}$ of the initial calibration.

Sample QA/QC

o   Method blanks must be analyzed for every 20 samples or a "case" (group of samples from a given site over a given time period and assigned a unique "case" number), whichever is less, for each matrix in a case, and for each level of concentration. Common phthalate esters must not appear in the blank at greater than five times the CRDL. All other semi-volatile target compounds must not be present in the blank at greater than the CRDL.

o   Surrogate compounds (see Table 4) are added to each sample prior to sample preparation at designated concentrations to determine if there is a problem caused by sample preparation, analysis, or matrix. If surrogate recoveries do not meet contractually required recovery windows, samples must be re-extracted and re-analyzed.

o   Matrix spikes are added to a given sample in duplicate for each matrix and concentration level   in a case.   Matrix spike recoveries provide some information on the suitability of the method for the sample matrix but should not be used to determine data useability for other than the given sample unless other information is taken into account. Matrix spike recovery information would ideally provide method precision information (its primary objective historically). However, the inability to exercise precision in sampling typical Superfund samples (largely non-homogenous and difficult to homogenize) has made it impractical to define recovery windows that are contractually required. However, there are recommended performance based QC limits for both relative percent difference (RPD) and percent recoveries of the matrix spike compounds. If a large deviation from these guidelines occurs, perhaps the method is unsuitable for the given maxtix or there is a laboratory bias that must be monitored and addressed if chronic.

o   The QA/QC data for semi-volatiles from all CLP laboratories is constantly monitored and evaluated by EMSL/LV. It is discussed in detail in a separate article in this STP (6).

CONCLUSION

The use of rigid analytical protocols, intense QA/QC and data reporting uniformity has successfully shown that the CLP can and does produce high quality data that can be interpreted and translated into information for use in remedial actions and/or in enforcement of the "liability" provisions of CERCLA when "responsible parties" are identified.

REFERENCES

(1)   Environmental Protection Agency Solicitation IFB WA 85-J664
(2)   Federal Register, Monday, December 3, 1979
(3)   Federal Register, Friday, October 26, 1984
(4)   Longbottem, J.E., and J.J. Lichtenberg, Ed., in Methods for Organic Chemical Analysis of Municipal and Industrial Wastewater, EPA-600/4-82-057, July 1982.
(5)   Fisk, J.F., Haeberer, A.M., and Kovell, S.P., Spectra, Volume 10, Number 3.
(6)   Wolff, J.S., Homsher, M.T., Flotard, R.D., and Pearson, J.G., "Semi-volatile Organic Analytical Methods Performance and Quality Control Considerations," Quality Control in Remedial Site Investigation: Hazardous and Industrial Solid Waste Testing, Fifth Volume, ASTM STP 925, C.L. Perket, Ed., American Society for Testing and Materials, 1986.

Jeffrey S. Wolff, Michael T. Homsher, Richard D. Flotard, and
J. G. Pearson

SEMI-VOLATILE ORGANIC ANALYTICAL METHODS PERFORMANCE AND QUALITY
CONTROL CONSIDERATIONS

REFERENCE: Wolff, J. S., Homsher, M. T., Flotard, R. D., and
Pearson, J. G., "Semi-Volatile Organic Analytical Methods
Performance and Quality Control Considerations," Quality
Control in Remedial Site Investigation: Hazardous and Indus-
trial Solid Waste Testing, Fifth Volume, ASTM STP 925, C. L.
Perkert, Ed., American Society for Testing and Materials,
Philadelphia, 1986.

ABSTRACT: The analysis of semi-volatile organic compounds in
hazardous waste samples for the U. S. Environmental Protection Agency's Superfund Program is done by selected contractor laboratories using methods specified in the EPA Invitation for Bid document WA-85-J664. The Contract Laboratory
Program maintains a data base containing information on
precision and accuracy for matrix and surrogate spiking of
samples. In addition, pre-award and quarterly blind quality
assurance sample results yield data for precision and accuracy. The precision, expressed as relative standard deviation
varies from 16% to 42% while the bias ranges from -59% to
+12%.

KEYWORDS: semi-volatile organic analysis, Contract Laboratory Program, CERCLA, quality assurance; quality control,
hazardous waste, gas chromatography, mass spectrometry, soil
matrix, water matrix, Superfund

The Environmental Protection Agency's Environmental Monitoring
Systems Laboratory-Las Vegas (EMSL-LV) is responsible for conducting a
quality assurance program in support of the Agency's Superfund Contract
Laboratory Program (CLP). For additional information on the CLP and

Mr. Wolff is a Senior Associate Scientist, Mr. Homsher is a
Scientific Supervisor and Dr. Flotard is a Principal Scientist at
Lockheed Engineering and Management Services Company, P.O. Box 15027,
Las Vegas, NV 89114; Mr. Pearson is Branch Chief of the Toxic and
Hazardous Waste Operations, Quality Assurance Division at the Environ-
mental Monitoring Systems laboratory, 944 E. Harmon Ave., Las Vegas,
NV 89109.

the EMSL-LV quality assurance program, see the papers in this volume by Fisk [1, 2], Kovell [3], Moore and Pearson [4], Flotard et al., [5] and the Users Guide to the Contract Laboratory Program [6].

The EMSL-LV quality assurance program provides analytical calibration standards and quality control (QC) material, maintains a quality assurance (QA) data base, and conducts audits of CLP data. Through these activities, EMSL-LV continuously evaluates and documents CLP laboratory and method performance. Analytical methods are dynamically evaluated and validated, i.e., the performance of the methods is continually assessed and periodically documented. Where these assessments indicate gaps in our knowledge about the performance of the methods, the EMSL-LV designs and conducts performance evaluation studies to fill these information gaps. If these assessments indicate that method performance is inadequate to meet the Agency's monitoring objectives, studies are conducted to gather the data necessary to improve method performance. The objective of this paper is to document the performance of the CLP analytical method for the analysis of semi-volatile organic compounds in water and soils/sediment.

APPLICATION AND METHOD DESCRIPTIONS

The semi-volatile organic analysis (BNA) procedures are used for the 65 target compounds in Table 1. If present in sufficient quantity in each sample, up to 20 additional non-target semi-volatile compounds are tentatively identified and their concentrations estimated using a forward search of the EPA/NIH library. The analysis of soil is divided into two methods, a low level and medium level. A single level water method incorporates a dilution step to allow for varying concentrations of the analytes.

Sample Screening

The screening of samples using capillary column gas chromatography with mass spectrometer or flame ionization detector (GC/MS or GC/FID) is recommended so that the appropriate low or medium level protocol may be used.

Low Level Soil

A 30-gram portion of sediment/soil sample is mixed with anhydrous sodium sulfate, surrogate spike solution is added (matrix spiking solution is added if applicable), and the sample is extracted using 1:1 methylene chloride:acetone (v:v) using an ultrasonic probe. If the low level screen is used, a portion of this dilute extract is concentrated fivefold and screened by GC/FID or GC/MS. If peaks are present at a concentration greater than 20,000 µg/kg, the extract is discarded and the sample is then prepared by the medium level method. If no peaks are present at a concentration greater than 20,000 µg/kg the extract is then concentrated and split into two fractions. An optional gel permeation column cleanup may be used before splitting the extract. One

TABLE 1--Semi-volatile HSL compounds and contract required
detection limits (CRDL).

Compound name	SPCCa or CCCb	Low Soil CRDL, µg/kg	Low Water CRDL, µg/L	CAS Number
Phenol	CCC	330	10	108-95-2
bis(2-Chloroethyl)ether		330	10	111-44-4
2-Chlorophenol		330	10	95-57-8
1,3-Dichlorobenzene		330	10	541-73-1
1,4-Dichlorobenzene	CCC	330	10	106-46-7
Benzyl alcohol		330	10	100-51-6
1,2-Dichlorobenzene		330	10	95-50-1
2-Methylphenol		330	10	95-48-7
bis(2-Chloroisopropyl)ether		330	10	39638-32-9
4-Methylphenol		330	10	106-44-5
N-Nitroso-di-n-propylamine	SPCC	330	10	621-64-7
Hexachloroethane		330	10	67-72-1
Nitrobenzene		330	10	98-95-3
Isophorone		330	10	78-59-1
2-Nitrophenol	CCC	330	10	88-75-5
2,4-Dimethylphenol		330	10	105-67-9
Benzoic acid		1,600	50	65-85-0
bis(2-Chloroethoxy)methane		330	10	111-91-1
2,4-Dichlorophenol		330	10	120-83-2
1,2,4-Trichlorobenzene		330	10	120-82-1
Naphthalene		330	10	91-20-3
4-Chloroaniline		330	10	106-47-8
Hexachlorobutadiene	CCC	330	10	87-68-3
4-Chloro-3-methylphenol	CCC	330	10	59-50-7
2-Methylnaphthalene		330	10	91-57-6
Hexachlorocyclopentadiene	SPCC	330	10	77-47-4
2,4,6-Trichlorophenol	CCC	330	10	88-06-2
2,4,5-Trichlorophenol		1,600	50	95-95-4
2-Chloronaphthalene		330	10	91-58-7
2-Nitroaniline		1,600	50	88-74-4
Dimethylphthalate		330	10	131-11-3
Acenaphthylene		330	10	208-96-8
3-Nitroaniline		1,600	50	99-09-2
Acenaphthene	CCC	330	10	83-32-9
2,4-Dinitrophenol	SPCC	1,600	50	51-28-5
4-Nitrophenol	SPCC	1,600	50	100-02-7
Dibenzofuran		330	10	132-64-9
2,4-Dinitrotoluene		330	10	121-14-2
2,6-Dinitrotoluene		330	10	606-20-2
Diethylphthalate		330	10	84-66-2
4-Chlorophenyl-phenylether		330	10	7005-72-3
Fluorene		330	10	86-73-7
4-Nitroaniline		1,600	50	100-01-6
4,6-Dinitro-2-methylphenol		1,600	50	534-52-1
N-Nitrosodiphenylamine	CCC	330	10	86-30-6
4-Bromophenyl-phenylether		330	10	101-55-3
Hexachlorobenzene		330	10	118-74-1
Pentachlorophenol	CCC	1,600	50	87-86-5
Phenanthrene		330	10	85-01-8
Anthracene		330	10	120-12-7

(continued)

TABLE 1--(Continued)

Compound name	SPCC[a] or CCC[b]	Low Soil CRDL, µg/kg	Low Water CRDL, µg/L	CAS Number
Di-n-butylphthalate		330	10	84-74-2
Fluoranthene	CCC	330	10	206-44-0
Pyrene		330	10	129-00-0
Butylbenzylphthalate		330	10	85-68-7
3,3'-Dichlorobenzidine		660	20	91-94-1
Benzo(a)anthracene		330	10	56-55-3
bis(2-Ethylhexyl)phthalate		330	10	117-81-7
Chrysene		330	10	218-01-9
Di-n-octylphthalate	CCC	330	10	117-84-0
Benzo(b)fluoranthene		330	10	205-99-2
Benzo(k)fluoranthene		330	10	207-08-9
Benzo(a)pyrene	CCC	330	10	50-32-8
Indeno(1,2,3-cd)pyrene		330	10	193-39-5
Dibenz(a,h)anthracene		330	10	53-70-3
Benzo(g,h,i)perylene		330	10	191-24-2

[a]CCC-Calibration Check Compound
[b]SPCC-System Performance Check Compound
Note: Medium soil/sediment contract required detection limits are 60
        times the individual low soil/sediment CRDL and medium water
        contract required detection limits are 100 times the individual
        low water CRDL.

fraction is used for analysis of the semi-volatile target compounds and
the other is used for pesticide analysis. The moisture content is
determined on a second aliquot of the same sample. Results of the soil
analysis are reported on a dry weight basis.

Medium Level Soil

    A 1-gram portion of sediment/soil sample is transferred to a vial,
mixed with 2 g of anhydrous sodium sulfate, and extracted by shaking
with methylene chloride. Surrogate and, if applicable, matrix spike
compounds are added immediately prior to extraction. The methylene
chloride extract is screened for extractable organics by GC/MS or
GC/FID. If organic compounds are detected by the screen, the methylene
chloride extract is analyzed using capillary column GC/MS analysis for
the target compounds in Table 1. A second aliquot of the same sample
is weighed and oven dried for the determination of moisture. Sample
results are reported on a dry weight basis.

Water

    A 1-liter water sample is transferred to a separatory funnel,
spiked with surrogate spiking solution (matrix spiking solution is
added if applicable), adjusted to pH 11 and extracted with three 60-mL
portions of methylene chloride. This extract is the base/neutral
fraction. The pH of the water is then readjusted to a pH 2 and re-
extracted with methylene chloride in a like manner to yield the acid
fraction. The extracts, treated individually, are dried over anhydrous
sodium sulfate and reduced in volume to 1 mL using a Kuderna-Danish

evaporative concentrator. A continuous liquid-liquid extractor may
be used in place of the separatory funnel if problems with emulsion
formation are anticipated. An optional GC screen may be performed on
the extracts. Capillary GC/FID analysis is used to estimate dilution.
The sample extracts are injected and compared to the results from a
50 ng injection of phenanthrene. Conditions are selected to give
approximately 50% of full scale deflection on GC for phenanthrene. The
sample dilution, if required, is used to bring sample peaks to between
50% and 100% of full scale under the same instrument conditions. The
internal standards are added to the extract immediately prior to analysis by GC/MS. Identification and quantification of target compounds is
achieved by matching both ion intensities and relative retention times
to known standard solutions using the internal standard method.

QUALITY CONTROL REQUIREMENTS

Introduction

       The quality control program is structured to provide consistent
results of known and documented quality. The program therefore places
stringent quality control requirements on all laboratories performing
sample analyses. Sample data packages contain documentation of a
series of QC operations that allow an experienced chemist to determine
the quality of the data and its applicability to each sampling effort.
In addition, laboratory contracts contain provisions for sample re-
analysis if and when specified QC criteria are not met by the contract
laboratory.

       Quality control requirements include: GC/MS instrument tune and
mass calibration, system performance checks, continuing calibration
checks, method blank analysis, internal standard area and retention
time monitoring, matrix spike/duplicates, surrogate spikes, criteria
for qualitative identification. A description of the quality control
requirements has been discussed in an earlier paper by Fisk [1] and is
discussed in detail in appendix E of the IFB WA-85-J664 [7].

GC/MS Instrument Tuning and Mass Calibration

       A 50-ng injection of DFTPP is used to establish that the GC/MS
meets standard mass spectral abundance criteria. The criteria must be
demonstrated daily or for each 12-hour period of instrument operation,
whichever is more frequent, before samples can be analyzed.

Initial Calibration

       Prior to the analysis of samples and after tuning criteria have
been met, the GC/MS system must be initially calibrated at five concentrations to determine the linearity of response utilizing target compound standards. The required standard amounts are 20, 50, 80, 120
and 160 ng per analyte. Response factors (RF) are then calculated
for each compound at each concentration level using Equation 1. Using

the five response factors (RF), or four response factors for the nine compounds with low response at 20 ng, from the initial calibration, the percent relative standard deviation (%RSD) is then calculated for each Calibration Check Compound (CCC) using Equation 2. The percent relative standard deviation for each individual Calibration Check Compound must be less than 30%. A system performance check is then performed to ensure minimum average response factors are met before the calibration curve is used. For semi-volatile System Performance Check Compounds (SPCC) the minimum acceptable average response factor is 0.050. No samples may be analyzed until the above conditions are met.

$$\text{Response factor (RF)} = \frac{A_x}{A_{is}} \times \frac{C_{is}}{C_x} \qquad (1)$$

where

A_x = area of the characteristic ion for the compound to be measured,
A_{is} = area of the characteristic ion for the specific internal standard,
C_{is} = concentration of the internal standard (ng/μL), and
C_x = concentration of the compound to be measured (ng/μL).

$$\text{Percent RSD} = \frac{SD}{\overline{X}} \times 100 \qquad (2)$$

where

RSD = relative standard deviation,
SD = standard deviation of initial five response factors (per compound), and
$\overline{X}$ = the mean of initial five response factors (per compound).

$$\text{Standard Deviation (SD)} = \sqrt{\sum_{i=1}^{N} \frac{(x_i - \overline{X})^2}{N-1}} \qquad (3)$$

Continuing Calibration

A calibration standard containing all semi-volatile target compounds, including all required surrogates, must be performed each 12 hours during analysis of samples. A system performance check is then made by evaluating the response factors of System Performance Check Compounds (SPCC). The minimum response factor for semi-volatile SPCC is 0.050. If the SPCC criteria are met, a comparison of response factors is made for all compounds. After the system performance check is met, Calibration Check Compounds (CCC) are used to check the validity of the initial calibration. The percent difference is calculated using Equation 4. If the percent difference for each CCC is less than 25% the initial calibration is assumed to be valid. If this criterion is not met for any one compound, corrective action must be made. If the problem cannot be corrected or a source of the problem found, a new initial calibration must be run before samples can be analyzed.

$$\text{Percent Difference} = \frac{\overline{RF}_i - RF_c}{\overline{RF}_i} \times 100 \qquad (4)$$

where

$\overline{RF}_i$ = average response factor from initial calibration, and
RF_c = response factor from current verification check standard.

Method Blank

A method blank must be performed for each batch of samples, for every 20 samples of similar concentration and matrix, or for each extraction procedure, whichever is more frequent. The method blank associated with a specific set or group of samples must be analyzed on each GC/MS system used to analyze that specific group or set of samples. A method blank is a volume of reagent water or a purified solid matrix carried through the entire analytical procedure. Its volume or weight must be approximately the size of the actual sample. A method blank for semi-volatile analysis must contain no greater than five times the Contract Required Detection Limit (CRDL) of common phthalate esters. For all other target compounds, the method blank must contain less than the CRDL of any single analyte. If criteria are not met, the system is out of control and corrective measures must be taken. All samples associated with a contaminated method blank must be re-extracted and re-analyzed.

Internal Standards

The internal standard method is used to quantify target compounds. Relative retention times of the target compounds are compared to the retention time of the nearest internal standard given in Table 4 and used to monitor changes in the operation of the gas chromatograph and mass spectrometer.

Internal standard responses and retention times in all samples should be evaluated during or immediately after data acquisition. If the retention time for any internal standard changes by more than 30 seconds, the chromatographic system is inspected for malfunctions and corrections are made. If the extracted ion current profile (EICP) area for any internal standard changes by more than a factor of two (-50% to 100%) from the latest daily calibration standard analysis, the mass spectrometric system is inspected for malfunction and corrections are made. Retention time and EICP area records are maintained in the form of control charts. All samples analyzed while the system was malfunctioning must be re-analyzed.

A matrix spike/matrix spike duplicate analysis must be performed once for each batch of samples, or with every 20 samples of similar concentration, whichever is more frequent. Relative percent difference and percent recovery are reported for each matrix spike sample. The quality control limits in the method are advisory only. Matrix spike information is used by the EPA to evaluate the long term precision of the analytical method.

TABLE 2--Internal standard compounds.

Compound name	Amount, ng/μL	CAS number
1,4-Dichlorobenzene-d_4	40	106-46-7
Naphthalene-d_8	40	1146-65-2
Acenapthene-d_8	40	13067-26-2
Phenanthrene-d_8	40	1517-22-2
Chrysene-d_{12}	40	1719-03-5
Perylene-d_{12}	40	1520-96-3

TABLE 3--Matrix spike compounds.

	Soil			
	Amount Low level, μg/kg		Amount Med level	Contract advisory recovery,
Compound name	GPC[a]	No GPC	μg/kg	%
1,2,4-Trichlorobenzene	6,670	3,300	100,000	38-107
Acenaphthene	6,670	3,300	100,000	31-137
2,4-Dinitrotoluene	6,670	3,300	100,000	28-89
Pyrene	6,670	3,300	100,000	35-142
N-Nitroso-di-n-propylamine	6 670	3,300	100,000	41-126
1,4-Dichlorobenzene	6,670	3,300	100,000	28-104
Pentachlorophenol	13,370	6,670	200,000	17-109
Phenol	13,370	6,670	200,000	26-90
2-Chlorophenol	13,370	6,670	200,000	25-102
4-Chloro-3-methylphenol	13,370	6,670	200,000	26-103
4-Nitrophenol	13,370	6,670	200,000	11-114

	Water	
	Amount, μg/L	Contract advisory recovery, %
1,2,4-Trichlorobenzene	100	39-98
Acenaphthene	100	46-118
2,4-Dinitrotoluene	100	24-96
Pyrene	100	26-127
N-Nitroso-di-n-propylamine	100	41-116
1,4-Dichlorobenzene	200	36-97
Pentachlorophenol	200	9-103
Phenol	200	12-89
2-Chlorophenol	200	27-123
4-Chloro-3-methylphenol	200	23-97
4-Nitrophenol	200	10-80

[a]GPC = Gel permeation chromatography.

Surrogate Spike Analysis

Surrogate standard determinations are performed on all samples and blanks. For a listing of surrogate spiking compounds, see Table 4. Surrogate spike recovery must be evaluated by determining whether or not the concentration, measured as percent recovery, falls inside the contract required recovery limits in Table 4. The laboratory must take corrective action by recalculating surrogate recoveries or reanalyzing a sample if any one surrogate compound recovery in either the base/

neutral or acid fraction is below 10% or recoveries of two surrogate compounds in either acid or base/neutral fractions are outside surrogate spike recovery limits.

TABLE 4--Surrogate spike compounds.

Compound name	Amount added- Low level,[a] μg	Amount added- Med level, μg	Contract required recovery limits, % soil	water	CAS number
Nitrobenzene-d5	50	50	23-120	35-114	4165-60-0
2-Fluorobiphenyl	50	50	30-115	43-116	321-60-8
p-Terphenyl-d14	50	50	18-137	33-141	1718-51-0
Phenol-d5	100	100	24-113	10-94	13127-88-3
2-Fluorophenol	100	100	25-121	21-100	367-12-4
2,4,6-Tribromophenol	100	100	19-122	10-123	118-79-6

[a]Note amount in sample extract at the time of injection (before any optional dilutions).

PERFORMANCE DATA

Data Base

An extensive QA/QC data base is maintained on the Agency's computer at Research Triangle Park, North Carolina. It contains information from each sample submitted for analysis. Data for instrument tuning and mass calibration, initial and continuing calibration, matrix spike and matrix spike duplicate, surrogate spike, and method blank results are collected. The principal use of the data base is for the production of monthly laboratory performance monitoring reports on exceptions to the criteria for the method. It also serves to provide data for setting future method performance criteria and for dynamic method validation. The results which follow were derived from the data base which contains data from samples analyzed by 35 CLP laboratories during the period December 1, 1984 to November 5, 1985.

Intralaboratory Precision

Intralaboratory precision for the semi-volatile method may be estimated from matrix spike/matrix spike duplicate sample data shown in Table 5. Twelve of the 65 compounds in the semi-volatile target compound list are added to duplicate matrix samples. The RPD at the 85th percentile had been chosen at a 1984 CLP method caucus to be used for determining advisory control limits for the matrix spike duplicate samples. Analysis of the data is complicated by some laboratories having either spiked at incorrect levels or having reported the data in an incorrect manner. All results are included in the data base, but it is not currently possible to reject such data although this process will be implemented shortly. After correct spiking and reporting levels are implemented, sample homogeneity and verified matrix effects can be addressed. In an attempt to remove non-representative results, the data surveyed have been limited in the water matrix to ranges of mean recovery from 50% to 120% of the contract specified spike of 100 µg per compound for the base neutral compounds. Similarly, for acids in water matrix, the acceptable range is from 40% to 120% of the 200 µg per compound contract specified spiking amount. In the soil/sediment matrix, the mean recoveries in the range of 50% to 120% of the contract defined spike amount for base neutrals (i.e., mean recoveries from 25 to 60 µg for a 50 µg spike) and 40% to 240% of the contract defined spike amount for acids (i.e., mean recoveries from 40 to 240 µg for a 100 µg spike) were used. A significant number of the water samples had been spiked with twice the amount of matrix spike acid compounds specified in the contract.

TABLE 5--Intralaboratory precision for semi-volatile compounds.

Compound name	N[a]	Mean %RSD	%RSD at 85th percentile	%RSD at method advisory limit
Soil				
1,2,4-Trichlorobenzene	352	9	17	20
Acenaphthene	345	10	14	22
2,4-Dinitrotoluene	327	11	21	27
Di-n-butylphthalate	298	12	21	28
Pyrene	320	10	20	22
N-Nitroso-di-n-propylamine	344	11	20	27
1,4-Dichlorobenzene	354	9	19	20
Pentachlorophenol	327	15	27	35
Phenol	446	10	18	30
2-Chlorophenol	455	10	19	28
4-Chloro-3-methylphenol	438	11	20	30
4-Nitrophenol	259	18	33	35
Water				
1,2,4-Trichlorobenzene	240	8	14	20
Acenapthene	270	8	13	22
2,4-Dinitrotoluene	217	11	23	27
Di-n-butylphthalate	201	11	19	28
Pyrene	307	12	20	22
N-Nitroso-di-n-propylamine	271	11	19	27
1,4-Dichlorobenzene	208	7	13	20
Pentachlorophenol	357	13	22	35
Phenol	222	11	19	30
2-Chlorophenol	460	9	15	28
4-Chloro-3-methylphenol	388	11	19	30
4-Nitrophenol	240	14	28	35

[a]N = Number of duplicate samples in the data set.

The laboratories report matrix spike results as relative percent difference (RPD) which is converted to the more familiar relative standard deviation (RSD) using Equation 5. A set of advisory quality control limits has been established, based upon the past performance of the method. The RPD (or corresponding RSD) at the 85th percentile has been chosen as the control limit for matrix spikes. The CLP data base has been used to obtain these values.

RPD (relative percent difference) values are calculated between the matrix spike and matrix spike duplicate. The RPD for each component is calculated using Equation 5.

$$RPD = \frac{|D_1 - D_2|}{(D_1 + D_2)/2} \times 100 = \frac{\%RSD}{\sqrt{2}} \qquad (5)$$

where

RPD = relative percent difference,
D_1 = first sample result,
D_2 = second sample result, and
$\%RSD$ = percent relative standard deviation.

Interlaboratory Precision

Interlaboratory precision may be obtained from three sources, surrogate spike, quarterly blind, and pre-award data. The six compounds used as surrogates are expected to have precision and accuracy similar to the target compounds they represent. Table 6 lists the results for the surrogate spikes in samples for the period from December 1, 1984 to November 5, 1985 for 35 laboratories. The data includes percent recoveries for all surrogate spikes except for outliers which amount to approximately 4% of the surrogate spike data base. Surrogate recoveries less than 20% of the true spike value or exceeding three standard deviations above the mean recovery for all data have been removed.

TABLE 6--Surrogate spike bias and precision.

Compound name	Soil N	Soil %Bias	Soil %RSD	Water N	Water %Bias	Water %RSD
Nitrobenzene-d5	8,626	-34	33	10,902	-28	26
2-Fluorobiphenyl	8,688	-28	30	10,959	-21	23
p-Terphenyl-d14	8,678	-22	36	10,886	-16	30
Phenol-d5	8,572	-34	32	9,535	-51	44
2-Fluorophenol	8,637	-30	33	10,030	-41	34
2,4,6-Tribromophenol	8,373	-31	36	10,111	-31	34

Source: CLP data base report for the period 12/01/84 to 11/05/85.

Method Bias

The bias or accuracy of the method may be estimated from the quarterly blind data submitted by the CLP laboratories as part of the Environmental Monitoring Systems Laboratory - Las Vegas quality assurance program, from surrogate spike and from pre-award data. Blind spiked water samples are shipped to participating CLP laboratories as part of the normal case load. The data in Table 7 covers six quarterly blind samples submitted from the 2nd quarter of FY84 to the 3rd quarter of FY85. The number of participating laboratories has grown from 25 to 37 during that interval. Each mean recovery value represents approximately 30 submissions, including duplicates, after removal of values outside the acceptability windows specified by the method. In general the greatest negative bias (i.e., lowest recoveries) are associated with phenolic compounds, chemicals that have a high water solubility. The recovery of other classes of compounds does not show any general trend.

Accuracy for soil/sediment analyses may be estimated from the target compounds analyzed as part of the pre-award program in 1985. Two sets of pre-award data were used to construct Table 7.

TABLE 7--Method bias and precision as determined by quarterly blind and pre-award data.

| Compound name | Water | | | Soil | | |
	Bias	N	Ave. mean %RSD	Bias	N	Ave. mean %RSD
bis(2-Chloroethyl)ether	-16	106	24			
2-Chlorophenol	-21	192	29			
1,3-Dichlorobenzene	-48	49	24			
1,4-Dichlorobenzene	-25	88	21	-51	10	27
1,2-Dichlorobenzene	-28	133	29			
2-Methylphenol	-30	145	29			
bis(2-Chloroisopropyl)ether	-22	47	25			
4-Methylphenol	-36	102	33			
N-Nitroso-di-n-propylamine	+0.3	146	31			
Nitrobenzene	-23	160	32	-48	9	21
Isophorone	-8	143	23	-47	19	24
2-Nitrophenol	-21	227	30	-36	10	35
bis(2-Chloroethoxy)methane	-2.6	101	34			
2,4-Dichlorophenol	-20	222	29	-59	10	31
1,2,4-Trichlorobenzene	-47	90	30	-43	20	28
Napthalene	-38	30	44			
4-Chloro-3-methylphenol	-32	186	26			
2,4,6-Trichlorophenol	-17	216	25			
2-Chloronapthalene	+3.4	52	24			
Acenapthene	-12	102	28			
2,4-Dinitrophenol	-23	22	24			
2,4-Dinitrotoluene	-33	58	34			
2,6-Dinitrotoluene	-48	86	25			
4-Chlorophenyl-phenylether	+12	103	34			
Fluorene	-24	89	25			
4,6-Dinitro-2-methylphenol	-13	32	30			
4-Bromophenyl-phenylether	-0.1	84	32			
Hexachlorobenzene	-42	49	36			
Pentachlorophenol	-24	42	31	-48	9	17
Phenanthrene	-28	47	21			
Fluoranthene	-15	122	42			
Benzo(b)fluoranthene	-10	34	39			
Benzo(a)pyrene	-29	38	42			
Pyrene				-15	10	25
2-Methylnaphthalene				-42	10	26
bis-(2-Ethylhexyl)phthalate				-2	10	33
Phenol				-27	10	38
Acenaphthylene				-27	9	26
Diethylphthalate				-20	9	16

Method Detection Limits

Method detection limits are highly dependent upon the type of matrix encountered in the sample, the presence of interferences, and upon any dilutions that may be required to bring the sample within the linear range of the instrument.

There is no direct data available leading to method detection limits, but they may be inferred from the instrument detection limits (IDL). The IDL is operationally defined as being three times the standard deviation for three analyses of standards having concentrations from three to five times the CRDL. The method detection limits obtained in this way are listed in Table 8. These probably represent the detection limits under the most optimum conditions.

The IDL provides an estimate of the minimum amount that is needed "on column" in the GC/MS to identify and quantify an analyte. It is a simple procedure to back calculate to how much an analyte would have to be in either a liter of water or kilogram of soil to give that amount on column in the GC.

TABLE 8--BNA method detection limits.[a]

Compound name	Average water, µg/L	Med[b] Soil, µg/kg
Phenol	7.2	3470
bis(2-Chloroethyl)ether	7.6	2840
2-Chlorophenol	6.5	3030
1,3-Dichlorobenzene	6.3	2860
1,4-Dichlorobenzene	5.1	2230
Benzyl alcohol	14.7	4080
1,2-Dichlorobenzene	4.8	2140
2-Methylphenol	6.6	2700
bis(2-Chloroisopropyl)ether	10.4	5320
4-Methylphenol	8.6	3800
N-Nitroso-di-n-propylamine	8.4	4560
Hexachloroethane	7.5	3470
Nitrobenzene	8.5	3490
Isophorone	7.5	2680
2-Nitrophenol	8.8	3580
2,4-Dimethylphenol	7.6	3220
Benzoic acid	49.4	10500
bis(2-Chloroethoxy)methane	8.9	2950
2,4-Dichlorophenol	6.0	2660
1,2,4-Trichlorobenzene	5.3	2400
Naphthalene	5.0	2050
4-Chloroaniline	35.1	5840
Hexachlorobutadiene	7.9	3100
4-Chloro-3-methylphenol	8.5	3490
2-Methylnaphthalene	17.1	2830
Hexachlorocyclopentadiene	20.2	4740
2,4,6-Trichlorophenol	8.6	4020
2,4,5-Trichlorophenol	15.9	4860
2-Chloronaphthalene	5.2	2280

(continued)

TABLE 8--(Continued)

Compound name	Average water, μg/L	Med[b] Soil, μg/kg
2-Nitroaniline	60.3	6010
Dimethyl phthalate	6.9	2810
Acenaphthylene	5.0	2130
3-Nitroaniline	65.3	9190
Acenaphthene	6.1	2300
2,4-Dinitrophenol	44.9	12400
4-Nitrophenol	19.8	9730
Dibenzofuran	6.6	2150
2,4-Dinitrotoluene	11.3	5250
2,6-Dinitrotoluene	7.3	3280
Diethylphthalate	7.3	3220
4-Chlorophenyl-phenylether	6.7	2820
Fluorene	6.5	3100
4-Nitroaniline	67.6	10600
4,6-Dinitro-2-methylphenol	28.9	7940
N-Nitrosodiphenylamine	7.2	3810
4-Bromophenyl-phenylether	5.8	2820
Hexachlorobenzene	7.5	3240
Pentachlorophenol	13.6	7200
Phenanthrene	5.5	2390
Anthracene	5.7	2520
Di-n-butylphthalate	7.9	3830
Fluoranthene	6.9	3300
Pyrene	7.7	3250
Butylbenzylphthalate	8.0	4160
3,3'-Dichlorobenzidine	15.7	6820
Benzo(a)anthracene	7.0	2800
bis(2-Ethylhexyl)phthalate	7.1	3470
Chrysene	6.5	2430
Di-n-octylphthalate	10.9	4600
Benzo(b)fluoranthene	8.7	3300
Benzo(k)fluoranthene	7.8	2810
Benzo(a)pyrene	9.1	3420
Indeo(1,2,3-cd)pyrene	17.7	10800
Dibenz(a,h)anthracene	17.8	10200
Benzo(g,h,i)perylene	14.7	7010

[a]Assumes a 1-μL injection into the GC/MS.
[b]Low soil MDLs are 1/30th of the Medium level soil MDLs.

CONCLUSIONS

The quality assurance program for the CLP has been designed to incorporate provisions for monitoring laboratory and method performance. An ongoing dynamic method validation makes changes in the method as problems are noted. A number of method changes have been made during the 5 years the program has been in existence and more are likely to follow.

The bias in the water matrix ranges from +12% to -48%, and for soil ranges from -2% to -59%. Compounds which are highly acidic or which are difficult to chromatograph tend to show the greatest bias. The bias for soil appears to be greater than that in water.

Deuterated surrogate compounds have a bias comparable to the corresponding target compounds in the quarterly blind and pre-award data. The range for interlaboratory precision in water, expressed as %RSD, is 21% to 42%. For soil the range for interlaboratory precision is 16% to 30% as determined by quarterly blind and pre-award data respectively. The precision, expressed as %RSD, for most compounds is approximately 30%.

The quality assurance program has been instrumental in ensuring that the CLP produces data of known and documented quality suitable for remedial and litigative actions.

ACKNOWLEDGMENTS

Although research described in this article has been funded wholly by the United States Environmental Protection Agency through contract number 68-03-3249, it has not been subjected to Agency review and therefore does not necessarily reflect the views of the Agency and no official endorsement should be inferred. Mention of trade names or commercial products does not constitute Agency endorsement of the product.

REFERENCES

[1] Fisk, J. F., "Semi-Volatile Organic Analytical Methods - General Description and Quality Control Considerations." Quality Control in Remedial Site Investigation: Hazardous and Industrial Solid Waste Testing, Fifth Volume, ASTM STP 925, C. L. Perket, Ed., American Society for Testing and Materials, Philadelphia, 1986.
[2] Fisk, J. F., "Volatile Organic Analytical Methods - General Description and Quality Control Considerations." Quality Control in Remedial Site Investigation: Hazardous and Industrial Solid Waste Testing, Fifth Volume, ASTM STP 925, C. L. Perket, Ed., American Society for Testing and Materials, Philadelphia, 1986.
[3] Kovell, S. P., "Contract Laboratory Program - An Overview." Quality Control in Remedial Site Investigation: Hazardous and Industrial Solid Waste Testing, Fifth Volume, ASTM STP 925, C. L. Perket, Ed., American Society for Testing and Materials, Philadelphia, 1986.
[4] Moore, J. M. and Pearson, J. G., "Quality Assurance Support for the Superfund Contract Laboratory Program." Quality Control in Remedial Site Investigation: Hazardous and Industrial Solid Waste Testing, Fifth Volume, ASTM STP 925, C. L. Perket, Ed., American Society for Testing and Materials, Philadelphia, 1986.

[5]   Flotard, R. D., et al., "Quality Control in Remedial Site In-
      vestigation: Hazardous and Industrial Solid Waste Testing, Fifth
      Volume, ASTM STP 925, C. L. Perket, Ed., American Society for
      Testing and Materials, Philadelphia, 1986.
[6]   Users Guide to the Contract Laboratory Program, United States
      Environmental Protection Agency, Office of Emergency and Remedial
      Response, Washington, DC 20640, October 1984.
[7]   Invitation For Bid, Solicitation No. WA-85-J680, United States
      Environmental Protection Agency, 400 M Street SW, Washington, DC,
      20460, August 1, 1985.

Joan F. Fisk

VOLATILE ORGANIC ANALYTICAL METHODS - GENERAL DESCRIPTION AND QUALITY CONTROL CONSIDERATIONS.

REFERENCE: Fisk, J.F., "Volatile Organic Analytical Methods -General Description and Quality Control Considerations," Quality Control in Remedial Site Investigation: Hazardous and Industrial Solid Waste Testing, Fifth Volume, ASTM STP 925, C.L. Perket, Ed., American Society for Testing and Materials, 1986.

ABSTRACT: As a result of the enactment of the Comprehensive Environmental Response, Compensation and Liability Act (CERCLA) in 1980, and the subsequent delegation of its Authority by the President of the United States to the Administrator of the Environmental Protection Agency, the Contract Laboratory Program (CLP) has been developed to analyze "Superfund" samples in a broad-based manner in order to obtain the most information about these samples with a reasonable investment in time and money. The CLP uses gas chromatography/mass spectrometry (GC/MS) as its primary tool for analysis of samples for organic constituents. More explicitly, it employs purge and trap techniques prior to MS detection for analysis of samples for volatile compounds. An elaborate Quality Assurance/Quality Control (QA/QC) system is in place to guarantee that data generated by the CLP is of known quality and will hold up to the rigors of litigation in addition to providing an ongoing monitoring of CLP laboratories QA/QC data in order to guarantee their successful performance.

KEYWORDS: CERCLA, Contract Laboratory Program (CLP), Superfund, Gas Chromatography/Mass Spectrometry (GC/MS), Volatile Organics Analysis (VOA), purge and trap, Quality Assurance/Quality Control (QA/QC) surrogates, matrix spikes, internal standards, System Performance Check Compounds (SPCC), Calibration Check Compounds (CCC).

Joan Fisk is a Project Officer/Chemist with the Environmenal Protection Agency, Office of Solid Waste and Emergency Response, Office of Emergency and Remedial Response, Hazardous Response Support Division, Analytical Support Branch, 401 M Street, S.W., Washington, D.C., 20460.

INTRODUCTION

Public Law 96-510 entitled the Comprehensive Environmental Response, Compensation and Liability Act (CERCLA) was enacted in 1980 and bestowed upon the President of the United States the authority to effect the removal from the environment of any hazardous substance, pollutant, or contaminant in order to protect public health and welfare and the environment. The President delegated his authority to the Administrator of the Environmental Protection Agency (EPA) upon passage of CERCLA - or better known as - "Superfund" - and thus began the saga of the Contract Laboratory Program, i.e., CLP.

In order to quickly identify analytes in samples which are of a hazardous or polluting nature it is necessary to have a system in place to quickly analyze samples generated from the National Priority List (NPL) sites or (Superfund Sites) by routine standardized methods and to produce the data for these samples in a uniform manner which can clearly identify the compounds of concern.

Another major facet of CLP anlaysis is the need for a rigid Quality Assurance/Quality Control (QA/QC) program built into CLP procedures. EPA's Environmental Monitoring Systems Laboratory in Las Vegas (EMSL-LV) monitors the QA/QC of the CLP and maintains an extraordinary QA/QC data base which enables EMSL to identify problems within both CLP methods and CLP laboratories. In addition, the QA/QC requirements, that are an integral part of the protocols, put the stamp of authenticity on all data generated by CLP labs so that data is of known quality if and when it reaches a court of law and can serve as witness to liability for the violation of our environment.

The purpose of this paper is to describe the methods used to analyze Superfund samples for volatile compounds, the reasons for using a broad-based approach, the Quality Assurance/Quality Control requirements contained in the methods, and the results of this unique approach of the CLP.

The specific analyses described herein are for volatile organics analysis (VOA) using purge and trap Gas Chromatography (GC) techniques prior to GC separation of analytes and mass spectometry (MS) for detection of those analytes. For detailed protocols the reader should see the most recent version of CLP methods for organics analysis (1).

A brief historical profile follows, as well as a fairly explicit description of the protocols used for analyses of volatile compounds. The importance of the procedural QA/QC requirements in generating data of known quality by all CLP labs will be discussed.

BACKGROUND

Until October 1984, CLP methods for VOA were based on EPA Method 624 (2)(3)(4), with revisions over the years to make them more suitable for samples of varying and often extremely difficult matrices. In October 1984, the CLP modified all contracts to utilize protocols called the COP (Consensus Organics Protocol) which emerged through a series of caucuses with technical representation from EPA Regions, CLP

Contractors, EMSL-LV and EMSL-Cincinnati, and chaired by EPA Head-quarters Project Officers. While "Concensus" sometimes meant "who shouted the loudest", overall the acceptance of these methods has been universal.

SCOPE OF ANALYSIS

Since CERCLA defines a hazardous substance as any substance which may present substantial danger to public health, welfare, or the environment and includes substances referenced in the Water Pollution Control Act, the Solid Waste Disposal Act, the Clean Air Act, and the Toxic Substances Control Act, the number of possible compounds for identification is almost infinite in terms of real-time analysis. In order to get around a potentially forbidding task, the CLP has identified a list of compounds in Exhibit C of CLP Statements of Work which it has designated at the "Hazardous Substance List" (HSL). This is a misnomer and should be called the Target Compound List meaning the compounds which are routinely searched for during organics analysis. This list is dynamic and can be altered to include new compounds of concern or to delete compounds for which the protocols are deemed to be inappropriate (as determined by Performance based QA/QC). Following in Table 1 is the list of Volatile Target Compounds and their associated Contract Required Detection Limits (CRDLs).

To broaden the scope of the GC/MS analysis for VOA's the methods require a search of the EPA/NIH National Standard Reference Data System (commonly called the NBS library) to tentatively identify the spectra of up to 10 of the largest GC peaks in the reconstructed ion current (RIC) chromatograms for each sample.

SCREENING OF VOA SAMPLES

While it is not mandatory to screen CLP samples for VOA prior to GC/MS, it is beneficial to the laboratory to prevent swamping of instrumentation and subsequent downtime and to provide information on the appropriate aliquot of sample to be used for purge and trap GC/MS for optimal detection within the linear range of the instrument.

Following is a summary of the recommended screening procedures for volatile samples encompassing two approaches for interpretation.

Option A — A standard mixture (Standard mixture #1) containing benzene, toluene, ethyl benzene and xylene is used to calculate an approximate concentration of aromatics in a sample as determined by GC/FID analysis of hexadecane extracts of both water and soil/sediment samples.

Option B — A standard mixture (Standard mixture #2) containing n-nonane and n-dodecane is used to calculate a factor to determine the appropriate dilution for purge and trap of a water sample or to determine whether the low or medium purge and trap method for soil/sediment is required.

TABLE 1 -- VOA Target Compound List and
Contract Required Detection Limits (CRDL)

Volatiles	CAS Number	Contract Required Detection Limits Low Water[a] ug/L	Low Soil/Sediment[b] ug/Kg
1. Chloromethane	74-87-3	10	10
2. Bromomethane	74-83-9	10	10
3. Vinyl Chloride	75-01-4	10	10
4. Chloroethane	75-00-3	10	10
5. Methylene Chloride	75-09-2	5	5
6. Acetone	67-64-1	10	10
7. Carbon Disulfide	75-15-0	5	5
8. 1,1-Dichloroethene	75-35-4	5	5
9. 1,1-Dichloroethane	75-35-3	5	5
10. trans-1,2-Dichloroethene	156-60-5	5	5
11. Chloroform	67-66-3	5	5
12. 1,2-Dichloroethane	107-06-2	5	5
13. 2-Butanone	78-93-3	10	10
14. 1,1,1-Trichloroethane	71-55-6	5	5
15. Carbon Tetrachloride	56-23-5	5	5
16. Vinyl Acetate	108-05-4	10	10
17. Bromodichloromethane	75-27-4	5	5
18. 1,1,2,2-Tetrachloroethane	79-34-5	5	5
19. 1,2-Dichloropropane	78-87-5	5	5
20. trans-1,3-Dichloropropene	10061-02-6	5	5
21. Trichloroethene	79-01-6	5	5
22. Dibromochloromethane	124-48-1	5	5
23. 1,1,2-Trichloroethane	79-00-5	5	5
24. Benzene	71-43-2	5	5
25. cis-1,3-Dichloropropene	10061-01-5	5	5
26. 2-Chloroethyl Vinyl Ether	110-75-8	10	10
27. Bromoform	75-25-2	5	5
28. 2-Hexanone	591-78-6	10	10
29. 4-Methyl-2-pentanone	108-10-1	10	10
30. Tetrachloroethene	127-18-4	5	5
31. Toluene	108-88-3	5	5
32. Chlorobenzene	108-90-7	5	5
33. Ethyl Benzene	100-41-4	5	5
34. Styrene	100-42-5	5	5
35. Total Xylenes	5	5

[a] Medium Water Contract Required Detection Limits (CDRL) for Volatile Compounds are 100 times the individual Low Water CRDL.

[b] Medium Soil/Sediment Contract Required Detection Limits (CRDL) for Volatile Compounds are 100 times the individual Low Soil/Sediment CRDL.

The GC/FID is standardized externally with a mixture of Standard mixes #1 and 2. Chromatograms of hexadecane extracts of samples (40 ml of aqueous sample extracted with 2 ml of hexadecane or 10 gms of soil/sediment plus 40 ml of reagent water extracted with 2 ml of hexadecane) are compared with chromatograms of hexadecane extracts of reagent blanks and of the standard mixes.

The decisions which can be made by the GC/FID screen are as follows:

A. For aqueous samples:

1) If no peaks are present except those in the reagent blanks, analyze a 5 ml water sample by purge and trap GC/MS.

2) If peaks are present prior to the n-dodecane and the aromatics are distinguishable, follow Option A to determine appropriate dilution for purge and trap GC/MS.

3) If peaks are present prior to the n-dodecane but the aromatics absent or indistinguishable use Option B, such that if all peaks are less than or equal to 3% of the n-nonane, analyze a 5 ml water sample by purge and trap GC/MS. Any peaks greater than or equal to 3% of the n-nonane indicate the necessity to measure the peak height or area of the major peak and to calculate a dilution factor as follows:

$$\frac{\text{Dilution}}{\text{factor}} = \frac{\text{peak area of sample major peak}}{\text{peak area of n-nonane}} \times 50$$

(Note: The sample should not be diluted until just prior to purge and trap GC/MS analysis)

B. For soil/sediment samples:

1) If no peaks are present except those in the reagent blank, analyze a 5 gm sample by low level purge and trap GC/MS method for soil/sediment samples.

2) If peaks are present prior to the n-dodecane and the aromatics are distinguishable, follow Option A and the concentration information found in Table II (below) to determine whether to use the low or medium level method of GC/MS analysis.

3) If peaks are present prior to the n-dodecane but the aromatics are absent or indistinguishable, use Option B as follows: Calculate a factor using the formula

$$x \text{ factor} = \frac{\text{peak area of sample major peak}}{\text{peak area of n-nonane}}$$

TABLE 2 -- Determination of GC/MS
Purge and Trap Method

X-Factor	Analyzed by	Approximate Concentration Range
0-1.0	low level method	0-1000 ug/Kg
>1.0	medium level method	>1000 ug/Kg

It is important to recognize when using information from the VOA screen that the response of aromatic compounds to GC/FID is 20 times greater than that of halomethanes and 10 times greater than that of haloethanes.

ANALYSIS OF VOAs (5)

There are three procedures for analysis of purgeables (VOA), depending on the sample matrix. These procedures are based on the purge-and-trap technique in EPA method 624 (2)(3)(4). The three procedures are as follows:

(1) Water samples - An inert gas is bubbled through five ml samples (or a 5 ml aliquot of an appropriate dilution) contained in a purging chamber at ambient temperature. The purgeables are transferred from the aqueous phase to the vapor phase which is then swept onto an absorbent column (trap). The trap is then heated and backflushed with carrier gas, desorbing the purgeable compounds and carrying them onto the GC column for separation and subsequent detection by the mass spectrometer.

(2) Low level soil/sediment samples - An inert gas is bubbled through a mixture of 5 gms (or a more appropriate weight) of sample plus reagent water in a purge chamber at an elevated purge temperature. The purgeable compounds are transferred to the trap and the analysis proceeds as in the water sample method in (1) above.

(3) Medium level soil/sediment samples - A measured amount of soil is extracted with methanol. A calculated portion of the methanol extract is spiked into 5 ml of reagent water. An inert gas is bubbled through this mixture of the methanol extract of the sample and reagent water at ambient temperature, and the analysis proceeds as with the water sample method in (1) above.

Gas Chromatography and Mass Spectrometric Detection. A 6 ft x 2 mm i.d. column packed with 1% SP-1000 on Carbopak B is used for GC/MS analysis of the purgeables fraction. For detection, the mass spectrometer is scanned from 35 to 260 amu every seven seconds or less at 70 ev in the electron impact (EI) ionization mode. The mass spectrum for 4-bromofluorobenzene (BFB) must meet the criteria in Table 3 prior to the analysis of standards, blanks or samples.

TABLE 3 -- BFB Key Ions and Abundance Criteria

Mass	Ion Abundance Criteria
50	15.0-40.0 percent of the base peak
75	30.0-60.0 percent of the base peak
95	base peak, 100 percent relative abundance
96	5.0-9.0 percent of the base peak
173	less than 1.00 percent of the base peak
174	greater than 50.0 percent of the base peak
175	5.0-9.0 percent of mass 174
176	greater than 95.0 percent but less than 101.0 percent of mass 174
177	5.0-9.0 percent of mass 176

Qualitative Identification. There are three requirements for identification of a purgeable compound (in water, low level soil/sediment, or medium level soil/sediment samples). These requirements are as follows:

(1) All m/z's present in the standard mass spectrum at a relative intensity greater than 10% of the most abundant m/z in the spectrum (100%) must be present in the sample spectrum.

(2) The relative intensities of the m/z's that are greater than 10% of the most abundant m/z must agree within +/- 20% between the standard and sample spectrum.

(3) M/z's greater than 10% relative intensity in the sample spectrum but not in the standard spectrum must be considered and accounted for by the mass spectral interpretor.

Quantification. VOA target compounds are quantified using multiple internal standards, with the internal standard to be used designated as the one nearest in retention time to that of a given analyte, as given in Table 4. The extracted ion current profile (EICP) area at the characteristic m/z of a given target compound analyte is used to calculate its concentration. The characteristic m/z's are given in Tables 5 and 6.

The equations for determining analyte concentration are as follows:

(1) For low and medium level water samples:

$$\text{Concentration (ug/L)} = (A_x) \, (I_s)/(A_{is})(RF)(V_o)$$

where: A_x = EICP area of the analyte to be measured at its characteristic m/z

A_{is} = EICP area of the specified internal standard at its characteristic m/z

I_s = amount of internal standard added (ng)

V_o = volume of water purged (mL) (to account for dilutions)

TABLE 4 -- Volatile Internal Standards with Corresponding
Analytes Assigned for Quantitation

Bromochloromethane	**1,4-Difluorobenzene**
Chloromethane	2-Butanone
Bromomethane	1,1,1-Trichloroethane
Vinyl Chloride	Carbon Tetrachloride
Chloroethane	Vinyl Acetate
Methylene Chloride	Bromodichloromethane
Acetone	1,2-Dichloropropane
Carbon Disulfide	trans-1,3-Dichloropropene
1,1-Dichloroethene	Trichloroethene
1,1-Dichloroethane	Dibromochloromethane
trans-1,2-Dichloroethene	1,1,2-Trichloroethane
Chloroform	Benzene
1,2-Dichloroethane	cis-1,3-Dichloropropene
1,2-Dichloroethane-d$_4$(surr)[a]	2-Chloroethyl Vinyl Ether
	Bromoform

Chlorobenzene-d$_5$

2-Hexanone
4-Methyl-2-Pentanone
Tetrachloroethene
1,1,2,2-Tetrachloroethane
Toluene
Chlorobenzene
Ethylbenzene
Styrene
Total Xylenes
Bromofluorobenzene (surr)
Toluene-d$_8$ (surr)

[a] (surr) = surrogate compound

TABLE 5 -- Characteristic Ions for Volatile Target Compounds

Parameter	Primary Ion[a]	Secondary Ion(s)
Chloromethane	50	52
Bromomethane	94	96
Vinyl chloride	62	64
Chloroethane	64	66
Methylene chloride	84	49, 51, 86
Acetone	43	58
Carbon disulfide	76	78
1,1-Dichloroethene	96	61, 98
1,1-Dichloroethane	63	65, 83, 85, 98, 100
trans-1,2-Dichloroethene	96	61, 98
Chloroform	83	85
1,2-Dichloroethane	62	64, 100, 98
2-Butanone	72	57
1,1,1-Trichloroethane	97	99, 117, 119
Carbon tetrachloride	117	119, 121
Vinyl acetate	43	86
Bromodichloromethane	83	85, 129
1,1,2,2-Tetrachloroethane	83	85, 131, 133, 166
1,2-Dichloropropane	63	65, 114
trans-1,3-Dichloropropene	75	77
Trichloroethene	130	95, 97, 132
Dibromochloromethane	129	208, 206
1,1,2-Trichloroethane	97	83, 85, 99, 132, 134
Benzene	78	-
cis-1,3-Dichloropropene	75	77
2-Chloroethyl vinyl ether	63	65, 106
Bromoform	173	171, 175, 250, 252, 254, 256
2-Hexanone	43	58, 57, 100
4-Methyl-2-pentanone	43	58, 100
Tetrachloroethene	164	129, 131, 166
Toluene	92	91
Chlorobenzene	112	114
Ethyl benzene	106	91
Styrene	104	78, 103
Total xylenes	106	91

[a] The primary ion should be used unless interferences are present, in which case, a secondary ion may be used.

TABLE 6 -- Characteristic Ions for Surrogate and
Internal Standards for Volatile Organic Compounds

Compound	Primary Ion	Secondary Ion(s)
SURROGATE STANDARDS		
4-Bromofluorobenzene	95	174, 176
1,2-Dichloroethane d-4	65	102
Toluene d-8	98	70, 100
INTERNAL STANDARDS		
Bromochloromethane	128	49, 130, 51
1, 4-Difluorobenzene	114	63, 88
Chlorobenzene d-5	117	82, 119

(2) For Soil/Sediment Samples:

Medium Level:
Concentration (ug/kg) = $(A_x)(I_s)(V_t)/(A_{is})(RF)(V_i)(W_s)(D)$

Low Level:
Concentration (ug/kg) = $(A_x)(I_s)/(A_{is})(RF)(W_s)(D)$

where: A_x, I_s, and A_{is} are as above, and

V_t = volume of total extract (uL) (10,000 uL or a fraction of this volume when dilutions are made)

V_i = volume of extract added (uL) for purging

D = 100 - % of moisture (dry wt)

W_s = weight of sample extracted or purged (g)

The response factor (RF) is obtained from the appropriate daily standard analysis and is calculated from the equation:

$$RF = (A_x)(C_{is})/(A_{is})(C_x)$$

where: A_x = EICP area at the characteristic m/z of the analyte

A_{is} = EICP area at the characteristic m/z of the specified internal standard

C_{is} = concentration of the internal standard

C_x = concentration of the analyte in the standard

QUALITY ASSURANCE/QUALITY CONTROL

As mentioned in the Introduction, the level of QA/QC attached to the CLP method assumes that the data produced are of a known quality and can be related to EPA Data Quality Objectives (DQO's) which are qualitative and quantitative statements developed by data users to define the quality of the data needed for a particular data collection activity to support specific remedial decisions or regulatory actions. The QA/QC aspects may be divided into two categories, namely, instrument or system performance QA/QC and method performance or sample QA/QC.

Instrument QA/QC

o   Prior to analysis of any standards, blanks, or samples the ion criteria for 4-bromoflurobenzene (BFB) must be met as defined earlier in Table 3. The BFB tune must be successfully demonstrated every 12 hours of analysis time.

o   Five point initial calibrations at designated concentrations are required prior to any analysis of blanks or samples to define the dynamic range of the GC/MS system. Certain compounds have been selected as System Performance Check Compounds (SPCC) and must have a minimum average response factor ($\overline{RF}$) of 0.3 (except for that of bromoform which must be 0.25). Other compounds have been designated as Calibration Check Compounds (CCC) and must have a percent relative standard deviation (% RSD) of less than 30 to ensure the validity of the initial calibration. Only when these SPCC and CCC requirements are met is the instrument calibration considered valid.

o   After 12 hours of analysis, a continuing calibration (at a designated concentration) must be performed following a successful BFB tune. Again, the minimum RF allowed is 0.3 (0.25 for bromoform) for SPCCs. For CCCs the RFs must not deviate more than 25% from the $\overline{RF}$ of the initial calibration Sample QA/QC.

Sample QA/QC

o   Method blanks must be analyzed for every 20 samples or a "case" (group of samples from a given site over a given time period and assigned a unique "case" number), whichever is less, for each matrix in a case, and for each level of concentration. Common laboratory solvents (e.g., methylene chloride, acetone, and toluene) must not appear in the blank at greater than five times the CRDL. All other volatile target compounds must not be present in the blank at greater than CRDL.

o   Surrogate compounds (see Table 6) are added to each sample prior to analysis at designated concentrations to determine if there is a problem caused by sample preparation, analysis, or matrix.

o   If surrogate recoveries do not meet contractually required recovery windows, samples must be reanalyzed.

o   Matrix spikes are added to a given sample in duplicate for each matrix and concentration level in a case. Matrix spike recoveries provide some information on the suitability of the method for the sample matrix but should not be used to determine data useability for other than the given sample unless other information is taken into account. Matrix spike recovery information would ideally provide method precision information (its primary objective historically). However, the inability to exercise precision in sampling typical Superfund samples (largely non-homogeneous and difficult to homogenize) has made it impractical to define recovery windows that are contractually required. However, there are recommended performance based QC limits for both relative percent difference (RPD) and percent recoveries of the matrix spike compounds. If a large deviation from these guidelines occurs, perhaps the method is unsuitable for the given maxtix or there is a laboratory bias that must be monitored and addressed if chronic.

o   The QA/QC data for volatiles from all CLP laboratories is constantly monitored and evaluated by EMSL/LV and is discussed in detail in a separate article in this STP (6).

CONCLUSION

The use of rigid analytical protocols, intense QA/QC and data reporting uniformity has successfully shown that the CLP can and does produce high quality data that can be interpreted and translated into information for use in remedial actions and/or in enforcement of the "liability" provision of CERCLA when "responsible parties" are identified.

REFERENCES

(1) Environmental Protection Agency Solicitation IFB WA 85-J664
(2) Federal Register, Monday, December 3, 1979
(3) Federal Register, Friday, October 26, 1984
(4) Longbottom, J.E., and J.J. Lichtenberg, Ed. in Methods for Organic Chemical Analysis of Municipal and Industrial Wastewater, EPA-600/4-82-057, July 1982.
(5) Fisk, J.F., Haeberer, A.M., and Kovell, S.P., Spectra, Volume 10, Number 3.
(6) Flotard, R.D., Homsher, M.T., Wolff, J.S., and Moore, J.M., "Volatile Organic Analytical Methods - Performance and Quality Control Considerations," Quality Control in Remedial Site Investigation: Hazardous and Industrial Solid Waste Testing, Fifth Volume, ASTM STP 925, C.L. Perket, Ed., American Society for Testing and Materials, 1986.

Richard D. Flotard, Michael T. Homsher, Jeffrey S. Wolff, and John M. Moore

VOLATILE ORGANIC ANALYTICAL METHODS PERFORMANCE AND QUALITY CONTROL
CONSIDERATIONS

REFERENCE: Flotard, R. D., Homsher, M. T., Wolff, J. S., and
Moore, J. M., "Volatile Organic Analytical Methods - Perform-
ance and Quality Control Considerations, "Quality Control in
Remedial Site Investigation: Hazardous and Industrial Solid
Waste Testing, Fifth Volume, ASTM STP 925, C. L. Perket, Ed.,
American Society for Testing and Materials, Philadelphia,
1986.

ABSTRACT: The analysis of volatile organic compounds in
hazardous waste samples from the U.S. Environmental Protection Agency Superfund program is done by selected contractor
laboratories using methods specified in the EPA Invitation to
Bid document WA85-J664. Data for the precision and accuracy
of the method is obtained from the quality assurance data in
the Contract Laboratory Program database. Method precision
is obtained from the results of matrix and surrogate spiking
of the samples. Bias is obtained from the results of.sur-
rogate spiking, preaward samples, and quarterly blind quality
assurance samples. The precision, expressed as relative
standard deviation, varies from 8 to 16% while the bias
varies from +6.5 to -46%.

KEYWORDS: Superfund, Contract Laboratory Program, CERCLA,
volatile organic analysis, quality assurance, quality control,
hazardous waste, purge and trap, gas chromatography, mass
spectrometry, soil matrix, water matrix

    The Environmental Protection Agency's Environmental Monitoring
Systems Laboratory - Las Vegas (EMSL-LV) is responsible for conducting
a quality assurance program in support of the Agency's Superfund Contract Laboratory Program (CLP). For additional information on the CLP
and the EMSL-LV quality assurance program, see the papers in this

    Dr. Flotard is a Principal Scientist, Mr. Homsher is a Scientific
Supervisor and Mr. Wolff is a Senior Associate Scientist at Lockheed
Engineering and Management Services Company, P.O. Box 15027, Las Vegas,
NV 89114; Mr. Moore is Manager of Data Audits/On-site Evaluations
Program for the USEPA Quality Assurance Division at the Environmental
Monitoring Systems Laboratory, 944 E. Harmon Ave., Las Vegas, NV 89109.

volume by Fisk [1, 2], Kovell [3], Moore and Pearson [4], Wolff, et al. [5], and the Users Guide to the Contract Laboratory Program [6].

The EMSL-LV quality assurance program provides analytical calibration standards and quality control (QC) material, maintains a quality assurance (QA) database, and conducts audits of CLP data. Through these activities the EMSL-LV continuously evaluates and documents CLP laboratory and method performance. Analytical methods are dynamically evaluated and validated i.e., the performance of the methods is continually assessed and periodically documented. Where these assessments indicate gaps in our knowledge about the performance of the methods, the EMSL-LV designs and conducts performance evaluation studies to fill these information gaps. If these assessments indicate that method performance is inadequate to meet the Agency's monitoring objectives, studies are conducted to gather the data necessary to improve method performance. The objective of this paper is to document the performance of the CLP analytical method for the analysis of volatile organic compounds in water and soils/sediment.

APPLICATION AND METHOD DESCRIPTIONS

The volatile organic analysis (VOA) procedures [7] are used for the 35 target compounds in Table 1. If present in sufficient quantity in each sample, up to 10 additional non-target volatile compounds are tentatively identified and their concentrations estimated using a forward search of the EPA/NIH library. The analysis of soil is divided into two methods, a low level method for analyte concentrations ranging from the Contract Required Detection Limits (CRDL) in Table 1 to approximately 1,000 μg/kg, and a medium level method for analyte concentrations in excess of approximately 1,000 μg/kg. A single level water method incorporates a dilution step to allow for varying concentrations of the analytes.

Sample Screening

It is generally recommended that samples be screened using a gas chromatograph (GC) equipped with a flame ionization detector prior to actual analysis using gas chromatography with mass spectrometric detection (GC/MS). In the screening of soil, a 10 g aliquot of the wet sample is added to 40 mL of reagent water and shaken for one minute. The suspension is centrifuged, the aqueous portion is decanted, and the decantate is extracted by using 2 mL of hexadecane. For the screening of water samples, a 40 mL aliquot of water is extracted with 2 mL of hexadecane. In both cases, the hexadecane extracts are analyzed by flame ionization gas chromatography, and the analyte concentrations are estimated by comparison to either an aromatic standard or a nonane/dodecane standard.

TABLE 1--Contract required detection limits
(CRDL)[a] of target compounds.

Target compound name	SPCC[b] CCC[c]	Low soil CRDL, µg/kg	Low water CRDL, µg/L	CAS number
Chloromethane	SPCC	10	10	74-87-3
Bromomethane		10	10	74-83-9
Vinyl Chloride	CCC	10	10	75-01-4
Chloroethane		10	10	75-00-3
Methylene Chloride		5	5	75-09-2
Acetone		10	10	67-64-1
Carbon Disulfide		5	5	75-15-0
1,1-Dichloroethene	CCC	5	5	75-35-4
1,1-Dichloroethane	SPCC	5	5	75-35-3
Trans-1,2-Dichloroethene		5	5	156-60-5
Chloroform	CCC	5	5	67-66-3
1,2-Dichloroethane		5	5	107-06-2
2-Butanone		10	10	78-93-3
1,1,1-Trichloroethane		5	5	71-55-6
Carbon Tetrachloride		5	5	56-23-5
Vinyl Acetate		10	10	108-05-4
Bromodichloromethane		5	5	75-27-4
1,1,2,2-Tetrachloroethane	SPCC	5	5	79-34-5
1,2-Dichloropropane	CCC	5	5	78-87-5
Trans-1,3-Dichloropropene		5	5	10061-02-6
Trichloroethene		5	5	79-01-6
Dibromochloromethane		5	5	124-48-1
1,1,2-Trichloroethane		5	5	79-00-5
Benzene		5	5	71-43-2
Cis-1,3-Dichloropropene		5	5	10061-01-5
2-Chloroethyl Vinyl Ether		10	10	110-75-8
Bromoform	SPCC	5	5	75-25-2
4-Methyl-2-pentanone		10	10	108-10-1
2-Hexanone		10	10	591-78-6
Tetrachloroethene		5	5	127-18-4
Toluene	CCC	5	5	108-88-3
Chlorobenzene	SPCC	5	5	108-90-7
Ethyl Benzene	CCC	5	5	100-41-4
Styrene		5	5	100-42-5
Total Xylenes		5	5	N.A.

[a]CRDL values obtained from the IFB WA85-J664 [7].
[b]System Performance Check Compounds (SPCC) are used to check compound
instability and degradation in the GC/MS and to insure minimum average
response factors are met prior to the use of the calibration curve.
[c]Column Check Compounds (CCC) are used to check the validity of the
initial calibration.
Note: Medium soil and water CRDLs are 100 times the low level CRDLs.

Low Level Soil Method

A weighed soil sample, either 1.0 or 5.0 g, is placed in a specially designed purge vessel together with 5 mL of reagent water, the
internal standard spiking solution in Table 2, and the surrogate spiking solution in Table 4. If the sample is chosen for duplicate matrix
spike analysis, the matrix spiking solution listed in Table 3 is added.

TABLE 2--Internal standard compounds - amount transferred to purge vessel.

Compound name	Amount, ng	CAS number
Bromochloromethane	250	74-97-5
1,4-Difluorobenzene	250	540-36-3
Chlorobenzene-D5	250	3114-55-4

TABLE 3--Matrix spiking compounds.

Compound name	Amount added, ng	Soil QC recovery limit, %	Water QC recovery limit, %	CAS number
Chlorobenzene	250	60-133	75-130	108-90-7
Toluene	250	59-139	76-125	108-88-3
Benzene	250	66-142	76-127	71-43-2
1,1-Dichloroethene	250	59-172	61-145	75-35-4
Trichloroethene	250	62-137	71-120	79-01-6

TABLE 4--Surrogate spiking compounds.

Compound name	Amount added to purge vessel, ng	Contract required recovery limits Soil, %	Water, %	CAS number
Toluene-d8	250	88-110	81-117	2037-26-5
4-Bromofluorobenzene	250	86-115	74-121	460-00-4
1,2-Dichloroethane-d4	250	76-114	70-121	17060-07-0

The purge vessel is heated to 40° ± 1°C and purged with an inert gas for 12 minutes to remove the VOA compounds from the sample. Purgeable compounds are collected on a Tenax trap which is subsequently desorbed into the GC inlet. GC/MS analysis is performed on the sample, permitting identification and quantification of the VOA compounds. Results are reported on a wet weight basis. A second soil sample is oven dried for the determination of water content.

Medium Level Soil Method

A weighed 4.0 g sample is placed in a 15 mL vial together with 9.0 mL reagent methanol and 1 mL of surrogate spiking solution. The mixture is shaken for 2 minutes. A 1 mL aliquot of the methanol extract is removed and saved for analysis. An aliquot of the methanol extract, the volume determined by previous GC screen of the sample, is added to 4.9 mL of reagent water with added internal standard. Sufficient reagent methanol is added to the water such that sample aliquot plus reagent methanol totals 100 μL. The mixture is placed in the purge vessel and purged at ambient temperature for 12 minutes. The remainder of the analysis is the same as the low level description.

Low Level Water Method

The analysis of water employs the use of a purge and trap apparatus.

A 5 mL aliquot of the water sample is placed in the purge device together with surrogates, internal standards, and the matrix spike if
the sample is a matrix spike sample. An inert gas is bubbled through
the purge vessel at ambient temperatures for 12 minutes, efficiently
transferring the purgable compounds from the sample into the gas. The
vapors are swept through a Tenax column where the purgable compounds
are trapped. After purging has been completed, the column is heated
and backflushed into the inlet of a gas chromatograph (GC). The gas
chromatograph is temperature programmed to separate the purgables which
are then detected using a mass spectrometer.

Medium Level Water Method

The procedure for analyzing medium level water is identical to the
low level water method, except for a dilution of the water sample using
reagent water prior to the purge. This is done to bring the sample
to within the linear response range of the GC/MS.

QUALITY CONTROL REQUIREMENTS

Introduction

The CLP QC program is structured to provide consistent results of
known and documented quality. The program therefore places stringent
quality control requirements on all laboratories performing sample
analyses. Sample data packages contain documentation of a series of
QC operations that allow an experienced chemist to determine the quality of the data and its applicability to each sampling effort. In addition, laboratory contracts contain provisions for sample re-analysis
if and when specified QC criteria are not met by the contract laboratory.

Quality control requirements include: GC/MS instrument tune and
mass calibration, system performance checks, continuing calibration
checks, method blank analysis, internal standard area and retention
time monitoring, matrix spike/duplicates, and surrogate spikes. A
description of the quality control requirements has been discussed in
an earlier paper by Fisk [2] and is discussed in detail in appendix E
of the IFB WA-85-J664 [7].

GC/MS Instrument Tune and Mass Calibration

It must be demonstrated that the GC/MS instrument meets ion abundance criteria required by the method for a 50 ng injection of p-
Bromofluorobenzene (BFB) prior to the running of any samples or standards and at the end of each 12 hour period during instrument operation. Alternatively, 50 µg of BFB may be added to 5 mL of reagent
water and the purge and trap used.

System Performance Check and Initial Calibration

The GC/MS system must be calibrated using 20, 50, 100, 150 and 200 ng per analyte to determine the linearity of the response for the analytical system, prior to the initial analysis and when the analyst is unable to meet continuing calibration criteria. Average response factors (RF) and percent relative standard deviation (%RSD) must be calculated using Equations 1 and 2 respectively for each continuing calibration compound (CCC) and system performance check compound (SPCC). SPCCs are used to check for compound instability or degradation in the GC/MS. The minimum average response factor for an SPCC must be 0.300, except for bromoform which is 0.250. The relative standard deviation for CCCs may not exceed 30%.

$$\text{Response factor (RF)} = \frac{A_x}{A_{is}} \cdot \frac{C_{is}}{C_x} \tag{1}$$

$$\% \text{ Relative Standard Deviation (\%RSD)} = \frac{SD}{\overline{X}} \cdot 100 \tag{2}$$

$$\text{Standard Deviation (SD)} = \sqrt{\sum_{i=1}^{N} \frac{(x_i - \overline{X})^2}{N-1}} \tag{3}$$

RF = response factor
A_x = area of the characteristic ion for the compound being measured.
A_{is} = area of the characteristic ion for the contract specified internal standard.
C_{is} = concentration of the internal standard, ng/μL.
C_x = concentration of the compound to be measured, ng/μL.
%RSD = Relative standard deviation in %.
SD = Standard deviation of initial 5 response factors (per compound).
$\overline{X}$ = Mean of initial 5 response factors (per compound).

Continuing Calibration Check

The continuing calibration check is performed once each 12 hours of instrument operation, once per batch of samples, or once each 20 samples, whichever is more frequent. The same 50 μg/L standard used in the initial calibration is used for the continuing calibration check. Both SPCCs and CCCs are checked. The criteria which must be met are: the minimum RF for SPCCs is 0.300 (except 0.250 for Bromoform) and the maximum percent difference for the CCCs using Equation 4 is 25%. Meeting the criteria assures the continuing validity of the initial calibration and that compound degradation is not a problem.

$$\text{percent difference} = \frac{\overline{RF_i} - RF_c}{\overline{RF_i}} \cdot 100 \tag{4}$$

$\overline{RF}_i$ = average response factor for the initial calibration.
RF_c = current response factor for the compound.

Method Blank

A method blank is analyzed once each 12 hours, once per batch of samples, or once each 20 samples, whichever is more frequent. A method blank is a volume of reagent water or a purified solid matrix carried through the entire analytical process. Its volume or weight must be approximately the size of the actual samples. The results of method blank analysis are used to monitor laboratory practices and insure that samples are not contaminated as the result of poor laboratory practices. A method blank is considered to be contaminated if it contains more than the CRDL for any single target compound in Table 1 except for the common laboratory solvents (methylene chloride, acetone, and toluene) which have a limit of five times the CRDL. The laboratory must correct the problem and reextract and reanalyze all samples associated with a contaminated blank.

Internal Standards

The internal standard method is used to quantify target compounds in the CLP program. Relative retention times of the target compounds are compared to the retention time of the nearest internal standard given in Table 2 and used to monitor changes in the operation of the gas chromatograph and mass spectrometer. If the retention time of any internal standard changes by more than 30 seconds, the chromatographic system must be inspected for malfunctions and corrections made as required. If the extracted ion current profile area for any one internal standard differs from the area for the most recent (12 hour) calibration standard by more than a factor of two (-50% to +100%), then the GC/MS must be inspected and malfunctions corrected. All samples analyzed while the system was malfunctioning must be reanalyzed.

Matrix Spiking Procedure

Matrix spike samples are analyzed in duplicate, one pair for each batch of samples, for each matrix type analyzed, or for each group of 20 samples, whichever is more frequent. Relative percent difference and recovery values are reported for each matrix spike compound in Table 3. Quality control limits for RPD and for percent recovery are reported for each matrix spike compound. Quality control limits for RPD and for percent recovery found in the method [7] are advisory only. Matrix spike information is used by the EPA to evaluate the long term precision of the analytical method.

Surrogate Spiking Procedure

Surrogate spikes from the compounds in Table 4 are added to each sample prior to extraction. The resulting analysis must meet the minimum recovery specifications of the method, or the affected sample must be reextracted and reanalyzed.

PERFORMANCE DATA

Database

An extensive QA/QC database is maintained on the Agency's computer at Research Triangle Park, North Carolina. It contains information from each sample submitted for analysis to a CLP laboratory. Data for instrument tuning and mass calibration, initial and continuing calibration, matrix spike and matrix spike duplicate, surrogate spike, and method blank results are collected. The principal use of the database is for the production of monthly laboratory performance monitoring reports on exceptions to the criteria for the method. It also serves to provide data for setting future method performance criteria and for dynamic method validation. The results which follow were derived from the database which contains data from samples analyzed by 35 CLP laboratories during the period December 1, 1984 to November 5, 1985.

Intralaboratory Precision

The within-laboratory precision for the VOA method may be estimated using results from the matrix spike/matrix spike duplicate samples. Five representative compounds from the list of 35 target compounds are matrix spiking compounds. The laboratories report results as relative percent difference (RPD) which is convertable to a more familiar relative standard deviation (RSD) using Equation 5. The RPD at the 85th percentile had been chosen at a 1984 CLP method caucus to be used for determining advisory control limits for the matrix spike duplicate samples. Analysis of the data is complicated by some laboratories having either spiked at incorrect levels or having reported the data in an incorrect manner. All results are included in the data base, but it is not currently possible to reject such data although this process will be implemented shortly. After correct spiking and reporting levels are implemented, sample homogeneity and verified matrix effects can be addressed. In an attempt to remove non-representative results, the data surveyed have been limited mean recoveries from 50% to 120% of the contract specified spike of 50 µg per compound (i.e., 25 to 60 µg for a 50 µg spike).

$$RPD = \frac{|D_1 - D_2|}{(D_1 + D_2) \div 2} \cdot 100 = \frac{\%RSD}{\sqrt{2}} \qquad (5)$$

RPD = Relative percent difference.
D_1 = Primary sample result.
D_2 = Duplicate sample result.
%RSD = Percent relative standard deviation.

Interlaboratory Precision

The results of surrogate spiking of the samples yields information on the interlaboratory precision and the bias of the method. The compounds used are deuterated analogs of some of the target compounds and

TABLE 5--Intralaboratory precision estimated from matrix spike duplicate data.

| | | Soil | | |
| | | Mean | %RSD at 85th | %RSD at method |
Compound name	N	%RSD	percentile	advisory QC limit
1,1-Dichloroethene	353	8.9	16.1	15.6
Trichloroethene	438	7.8	13.5	17.0
Chlorobenzene	432	6.7	13.1	14.9
Toluene	376	7.4	14.0	14.9
Benzene	411	7.4	14.3	14.9

| | | Water | | |
| | | Mean | %RSD at 85th | %RSD at method |
	N	%RSD	percentile	advisory QC limit
1,1-Dichloroethene	591	5.3	9.1	9.9
Trichloroethene	729	5.2	8.7	9.9
Chlorobenzene	739	4.4	7.6	9.2
Toluene	739	5.0	8.3	9.2
Benzene	736	5.2	8.3	7.8

N   =   Number of duplicate samples in the data set.
%RSD =   Percent relative standard deviation.
Source: Matrix/matrix spike duplicate data from 35 CLP laboratories in samples submitted December 1, 1984 to March 15, 1985.

would be expected to have precision and accuracy identical to the target compounds they represent. Tables 6 and 7 list the results for the surrogate spikes in samples for the period December 1, 1984 to November 5, 1985 for 35 laboratories. The data include percent recoveries for all surrogate spikes except for outliers. Surrogate recoveries less than 20% of the true spike value or exceeding three standard deviations above the mean recovery for all data have been removed.

TABLE 6--Precision and bias for surrogate spiking compounds in soil.

| | All data | | | Outliers removed | | |
Compound name	N	Bias, %	%RSD	N	Bias, %	%RSD
Toluene-d_8	10399	+2.0	13.2	10293	+2.0	10.4
4-Bromofluorobenzene	10395	-1.7	15.9	10329	-2.0	12.1
1,2-Dichloroethane-d_4	10402	-3.2	18.1	10295	-3.8	12.7

TABLE 7--Precision and bias for surrogate spiking compounds in water.

| | All data | | | Outliers removed | | |
Compound name	N	Bias, %	%RSD	N	Bias, %	%RSD
Toluene-d_8	12689	-0.3	7.1	12643	-0.3	6.2
4-Bromofluorobenzene	12687	+0.4	8.3	12627	0.0	7.1
1,2-Dichloroethane-d_4	12685	-3.6	14.9	12636	-3.9	9.8

Method Bias

Data for method bias for volatile compounds in soil are very limited. The data for target compounds which are available for soil are from preaward analyses required of the laboratories seeking contracts in the CLP program. Excluded were data from laboratories which failed to be awarded contracts. The data in Table 8 for water were obtained from the results of the quarterly blind performance evaluation samples. Outliers were identified using the Grubb's test [8] and were excluded. The results for bias in Tables 6 and 7 are small when compared to the corresponding non-deuterated analog in Table 8. This indicates that surrogate recoveries may be optimistic, owing to the fact that spikes are added and then immediately extracted. Removal of outliers increases the bias while decreasing the %RSD because some outlier recoveries are in excess of 100%. Outliers account for approximately 1% of the data. The large positive bias for methylene chloride in Table 8 is probably the result of laboratory contamination of samples by methylene chloride used as a solvent in the laboratory.

TABLE 8--Precision and Bias for volatile organic analysis in soil and water.

	Soil			Water		
Compound name	Bias, %[a]	%RSD	N	Bias, %[b]	%RSD	N
Methylene chloride				+36.6	56	29
1,1-Dichloroethene				-26.3	20	64
1,1-Dichloroethane				-46.4	13	90
Trans-1,2-Dichloroethene				-21.7	31	68
Chloroform	-0.1	8.0	9	-21.1	12	126
1,2-Dichloroethane	+11.1	13.1	14	+2.4	13	104
1,1,1-Trichloroethane				-41.0	19	116
Carbon Tetrachloride				-32.1	12	105
1,1,2,2-Tetrachloroethane				-5.8	11	66
Bromodichloromethane				-13.0	19	116
1,2-Dichloropropane				-12.9	18	30
Trans-1,3-Dichloropropene				-41.2	31	53
Trichloroethene				-22.8	17	56
Dibromochloromethane	-12.0	35.0	9	-3.3	14	56
1,1,2-Trichloroethane				-7.0	11	38
Benzene	-10.3	32.1	9	-3.3	12	49
Cis-1,3-Dichloropropene				-35.5	22	53
Bromoform	-12.1	16.6	10	+6.5	16	127
2-Hexanone	-45.5	16.6	6			
Tetrachloroethene				-42.5	13	152
Toluene	+13.7	13.8	10	-23.3	14	38
Chlorobenzene	+13.2	21.2	10	-15.9	14	68
Ethyl Benzene				-31.9	4	49

[a]Bias for soil obtained from 1985 preaward results for IFB WA85-J005/006/175/176/177.
[b]Bias for water obtained from quarterly blind performance evaluation samples with 26 to 34 laboratories participating.
Note: Not all 35 VOA target compounds have been used in preaward or quarterly blind samples.

Method Detection Limits

The method detection limits (MDL) in Table 9 are highly dependent upon the type of matrix encountered in the sample, the presence of interferences, and upon any dilutions that may be required to bring the sample within the linear range of the instrument. There are no

direct data available leading to method detection limits, but they
may be estimated by calculation from the instrument detection limits
(IDL) provided by each laboratory as part of the data package submitted
by the CLP laboratories for each batch of samples. The IDL provides
an estimate of the minimum amount that is needed "on column" in the
GC/MS to identify and quantify an analyte. It is a simple procedure

TABLE 9--Estimated VOA detection limits.

Compound name	Low Soil, µg/kg	Med. Soil, µg/kg	Water, µg/L
Chloromethane	9.2	1160	9.2
Bromomethane	5.0	623	5.0
Vinyl Chloride	8.2	1070	8.2
Chloroethane	4.4	575	4.4
Methylene Chloride	3.7	449	3.7
Acetone	11.4	1450	11.4
Carbon Disulfide	5.8	751	5.8
1,1-Dichloroethene	4.1	527	4.1
1,1-Dichloroethane	3.8	489	3.8
Trans-1,2-Dichloroethene	3.6	468	3.6
Chloroform	2.9	384	2.9
1,2-Dichloroethane	2.8	3605	2.8
2-Butanone	8.5	1090	8.5
1,1,1-Trichloroethane	3.9	487	3.9
Carbon Tetrachloride	3.5	444	3.5
Vinyl Acetate	8.9	1140	8.9
Bromodichloromethane	3.4	453	3.4
1,2,2,2-Tetrachloroethane	4.3	551	4.3
1,2-Dichloropropane	3.1	393	3.1
Trans-1,3-Dichloropropene	3.2	414	3.2
Trichloroethene	3.1	393	3.1
Dibromochloromethane	3.2	409	3.2
1,1,2-Trichloroethane	4.3	544	4.3
Benzene	2.7	340	2.7
Cis-1,3-Dichloropropene	2.9	363	2.9
2-Chloroethyl Vinyl Ether	5.3	600	3.6
Bromoform	3.9	504	3.8
4-Methyl-2-pentanone	8.1	1040	8.1
2-Hexanone	7.6	974	7.6
Tetrachloroethene	4.1	526	4.1
Toluene	3.9	501	3.9
Chlorobenzene	3.3	435	3.3
Ethyl Benzene	5.2	672	5.2
Styrene	5.3	693	5.3
Total Xylenes	N.A.	N.A.	N.A.

Source: Calculated from instrument detection limits submitted by 24
laboratories from March to August 1985.

to back calculate how much of an analyte would have to be in either a
liter of water or kilogram of soil to give that amount on column in
the GC. Since the sample size for low water is 5 mL and that of low
soil is 5 g, the multiplication factor is identical for both cases,
and so are the MDLs. The IDL is operationally defined as being three

times the standard deviation for three analyses of a standards solution. This solution must contain the target compounds at concentrations in the range from 3 to 5 times the CRDL values in Table 1. These
probably estimate the detection limits under the most optimal conditions.

CONCLUSIONS

The quality assurance program for the CLP program has been designed
to incorporate provisions for monitoring laboratory and method performance. An ongoing dynamic method validation makes changes in the
method as problems are noted. Data are available for several years,
but changes in the method make comparisons difficult. A number of
method changes have been made during the five years the program has
been in existence and more are likely to follow.

Data for bias are available for few of the target compounds in the
soil matrix because only a few of the target compounds or the deuterated analogs are used for preaward soil spiking solutions or surrogate spike compounds, respectively. As additional laboratories com-
pete for preawards, more of the compounds will be used for that purpose, and additional data will become available. Matrix and surrogate
spiking data are derived from a substantial set of data.

Interlaboratory precision is available for 14% of the compounds
in the volatiles target compound list (5 of 35). In addition, 3 deuterated analogs of other compounds in the list are used as surrogate
spiking compounds. The precision for the non-deuterated compounds
may be inferred from the precision of the corresponding deuterated
compound.

The quality assurance program has been instrumental in ensuring
that the CLP produces data of known and documented quality suitable
for remedial and litigative actions.

ACKNOWLEDGEMENT

Although research described in this article has been funded wholly
by the United States Environmental Protection Agency under contract
number 68-03-3249 to Lockheed Engineering and Management Services Company, it has not been subjected to Agency review and therefore does
not necessarily reflect the views of the Agency, and no official endorsement should be inferred. Mention of trade names or commercial
products does not constitute Agency endorsement of the product.

The authors wish to acknowledge the assistance of Forest Garner,
Mary Flynn, Sean Weigand, Mike Kershaw and Henry Kerfoot in assembling
the data.

REFERENCES

[1] Fisk, J. F., "Semi-volatile Organic Analytical Methods - General Description and Quality Control Considerations." Quality Control in Remedial Site Investigation: Hazardous and Industrial Solid Waste Testing, Fifth Volume, ASTM STP 925, C. L. Perket, Ed., American Society for Testing and Materials, Philadelphia, 1986.

[2] Fisk, J. F., "Volatile Organic Analytical Methods - General Description and Quality Control Considerations." Quality Control in Remedial Site Investigation: Hazardous and Industrial Solid Waste Testing, Fifth Volume, ASTM STP 925, C. L. Perket, Ed., American Society for Testing and Materials, Philadelphia, 1986.

[3] Kovell, S. P., "Contract Laboratory Program - An Overview" Quality Control in Remedial Site Investigation: Hazardous and Industrial Solid Waste Testing, Fifth Volume, ASTM STP 925, C. L. Perket, Ed., American Society for Testing and Materials, Philadelphia, 1986.

[4] Moore, J. M., and J. G. Pearson, "Quality Assurance Support for the Superfund Contract Laboratory Program." Quality Control in Remedial Site Investigation: Hazardous and Industrial Solid Waste Testing, Fifth Volume, ASTM STP 925, C. L. Perket, Ed., American Society for Testing and Materials, Philadelphia, 1986.

[5] Wolff, J. S., Homsher, M. T., Flotard, R. D., and Pearson, J. D., "Semi-Volatile Organic Analytical Methods Performance and Quality Control Considerations." Quality Control in Remedial Site Investigation: Hazardous and Industrial Waste Testing, Fifth Volume, ASTM STP 925, C. L. Perket, Ed., American Society for Testing and Materials, Philadelphia, 1986.

[6] Users Guide to the Contract Laboratory Program, United States Environmental Protection Agency, Office of Emergency and Remedial Response, Washington, D.C. 20640, October 1984.

[7] Invitation For Bid, Solicitation No. WA-85-J664, United States Environmental Protection Agency, 400 M Street SW, Washington, D.C., 20460, August 1, 1985.

[8] Grubbs, F. E., "Sample Criteria for Testing Outlying Observations." Annals of Mathematics Statistics, Vol. 21, pp 27-58, March 1950.

PAUL J. MARSDEN, J. GARETH PEARSON, AND DAVID W. BOTTRELL

PESTICIDE ANALYTICAL METHODS - GENERAL DESCRIPTION AND
QUALITY CONTROL CONSIDERATIONS

REFERENCE: Marsden, P.J., Pearson, J.G. and D.W. Bottrell,
"Pesticide Analytical Methods - General Description and
Quality Control Consideration", Quality Control in Remedial
Site Investigation: Hazardous and Industrial Solid Waste
Testing, Fifth Volume, ASTM STP 925, C.L. Perket, Ed.,
American Society for Testing and Materials, Philadelphia, 1986.

ABSTRACT: Environmental samples are analyzed for
organochlorine pesticides/PCB's as part of the Environmental
Protection Agency's (EPA) Superfund Contract Laboratory
Program (CLP). A new protocol is required because
laboratories have difficulty meeting the present data
acceptability criteria and because the surrogate now used is
more chemically labile than the analytes determined with the
protocol. The new pesticide/PCB protocol specifies an
ultrasonic extraction procedure for soil and either continuous
liquid-liquid or separatory funnel extraction of water. It
describes clean-up procedures using gel permeation
chromatography, adsorbtion columns and mercury, copper or
tetrabutylammonium sulfate for sulfur removal. The protocol
requires a dual gas chromatography (GC) column quantitation of
17 single component pesticides and nine multicomponent
pesticides/PCB's using electron capture detection (ECD). The
quality assurance/quality control (QA/QC) requirements include
12-hour GC performance checks, instrument and method blank
requirements, matrix spike/matrix spike duplicate analyses
every 20 samples and periodic performance checks of clean-up
techniques. Validation of the protocol is being accomplished
according to EPA guidelines for determining ruggedness,
precision, bias, detection limits, and sensitivity.

KEYWORDS: Aroclor, CERCLA, Contract Laboratory Program (CLP),
Electron Capture Detector (ECD), Gas Chromatography (GC), Gel
Permeation Chromatography (GPC), hazardous waste, method
validation, organochlorine pesticides, PCB's, Quality
Assurance/Quality Control (QA/QC), Superfund.

Paul Marsden is a Staff Scientist with the S-CUBED Chemistry
Group, P.O. Box 1620, La Jolla. S-CUBED is a division of Maxwell
Laboratories. Gareth Pearson is a Branch Chief of the Toxics and
Hazardous Waste Operations Branch with the Environmental Protection

Agency, Environmental Monitoring and Systems Laboratory (EMSL-LV), 944 E. Harmon, Las Vegas, NV 89114. Dave Bottrell is a Chemist of the Hazardous Waste Operations Branch at the EMSL-LV.

INTRODUCTION

In 1980 the U.S. Environmental Protection Agency (EPA) was charged with determining the extent of environmental contamination by hazardous substances and with removing pollutants from the environment under the Comprehensive Environmental Response, Compensation and Liability Act (CERCLA or "Superfund"). Because of the magnitude of the task, the EPA sought the expertise and the support of independent laboratories to perform the large volume of analyses required under CERCLA. The mechanism by which this was accomplished was to set up the Contract Lab Program (CLP). CLP laboratories perform environmental analyses using specific protocols with rigid quality assurance/quality control (QA/QC) requirements for a list of compounds in the hazardous substance list (HSL). The HSL is divided into organic analytes [1, 2] and inorganic analytes [3]. Twenty-six of the organic HSL analytes are organochlorine pesticides and PCB's and are listed in Table 1. Because of particular concern over those compounds, they are determined at lower concentration levels than the other 100 HSL organics by using gas chromatography with electron capture detection (GC/ECD).

The original CLP pesticide/PCB protocol [4] was developed as a consensus method by EPA and contract employees from existing procedures such as EPA Method 608 [5], Method 8080 [6] from SW846 and procedures used by regional EPA [7] or state laboratories. Problems that were observed by the CLP laboratories or by EPA data audit personnel were resolved by modifying the protocol during CLP caucuses or by performing limited method development studies at the Environmental Monitoring Systems Laboratory in Las Vegas (EMSL-LV). As a result of years of such corrections, the present protocol has become a method mosaic derived from many sources. Rather than continue to modify the method, the EPA decided to write and validate a new method as part of their overall quality assurance effort for CERCLA analyses. This paper describes this new proposed pesticide/ PCB protocol.

REQUIREMENTS OF THE METHOD

The analytical procedure used for organochlorine pesticides and PCB's for the CLP must be suitable to quantitate the 17 single component pesticides and the nine multicomponent pesticides/PCB's of the HSL list in soil/sediment and water samples at the detection limits given in Table 1. The procedure must be applicable for use in a production mode wherein large numbers of samples are analyzed by a number of different laboratories. Rigid QA/QC requirements must be built into the procedure so the data collected by the different CLP laboratories can be included in a data base on the chemical contamination of the environment. In addition, all data generated by CLP labs with the procedure must be of sufficiently high quality to be legally defensible.

TABLE 1--HSL Pesticides/PCB's

Pesticides/PCB's	CAS Number	Detection Limits* Low Water µg/L	Detection Limits* Low Soil/Sediment** µg/Kg
alpha-BHC	319-84-6	0.05	2.0
beta-BHC	319-85-7	0.05	2.0
delta-BHC	319-86-8	0.05	2.0
gamma-BHC (Lindane)	58-89-9	0.05	2.0
Heptachlor	76-44-8	0.05	2.0
Aldrin	309-00-2	0.05	2.0
Heptachlor Epoxide	1024-57-3	0.05	2.0
Endosulfan I	959-98-8	0.05	2.0
Dieldrin	60-57-1	0.10	4.0
4,4'-DDE	72-55-9	0.10	4.0
Endrin	72-20-8	0.10	4.0
Endosulfan II	33213-65-9	0.10	4.0
4,4'-DDD	72-54-8	0.10	4.0
Endosulfan Sulfate	1031-07-8	0.10	4.0
4,4'-DDT	50-29-3	0.10	4.0
Endrin Ketone	53494-70-5	0.10	4.0
4,4'-Methoxychlor	72-43-5	0.5	20.0
Chlordane (technical)	12789-03-6	0.5	20.0
Toxaphene	8001-35-2	1.0	40.0
AROCLOR-1016	12674-11-2	0.5	20.0
AROCLOR-1221	11104-28-2	0.5	20.0
AROCLOR-1232	11141-16-5	0.5	20.0
AROCLOR-1242	53469-21-9	0.5	20.0
AROCLOR-1248	12672-29-6	0.5	20.0
AROCLOR-1254	11097-69-1	1.0	40.0
AROCLOR-1260	11096-82-5	1.0	40.0

* Specific detection limits are highly matrix dependent. The detection limits listed herein are provided for guidance and may not always be achievable.

** Detection limits listed for soil/sediment are based on wet weight. The detection limits calculated by the laboratory for soil/sediment, calculated on dry weight basis, as required by the protocol, will be higher.

PROBLEMS WITH THE PRESENT PROTOCOL

The major problems with the pesticide/PCB protocol now used in the program are: variable recoveries of the surrogate, poor performance of the sample cleanup procedures on some samples and inability of some laboratories to meet the periodic GC performance requirements for analyte breakdown and detector linearity. The specifics of the problems and their possible solutions are detailed in this section.

The protocol currently requires that dibutylchlorendate (Figure 1) be added to each sample and to blanks as a surrogate in order to monitor pesticide recovery and system chromatography. Dibutylchlorendate (DBC) was an unfortunate choice as a surrogate because unlike the analytes determined by the method, it is subject to ester hydrolysis. Dibutylchlorendate also co-elutes with dioctyl phthalate interferences which can cause the appearance of high surrogate recovery in some samples (the data audit group at the EMSL/LV has observed reported dibutylchlorendate recoveries from 0 to over 3,000 percent). The suitability of isodrin, octachlorobiphenyl isomers or hexabromobenzene as surrogates were investigated because of their similarity to the analytes determined by this method. Hexabromobenzene was selected as a surrogate because it gave reproducible recoveries and eluted in an area of the chromatograms relatively free from matrix interference. DBC has been retained in the method as an internal standard for establishing relative retention times.

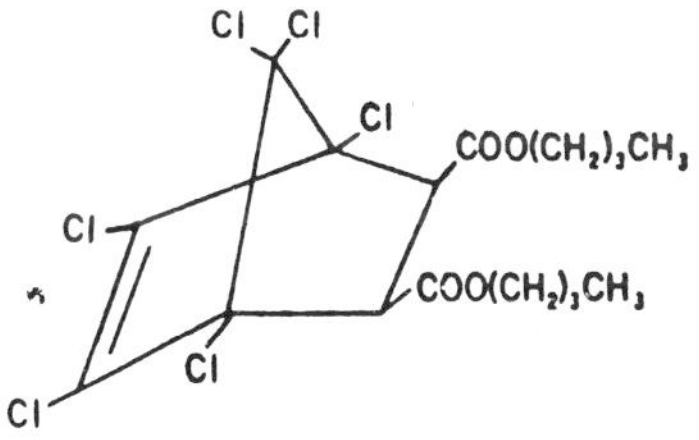

Figure 1. Dibutylchlorendate.

A second area where problems with the present protocol have been observed are in the cleanup procedures. Gel permeation chromatography (GPC) is now an optional step in sample preparation. Too few laboratories use the technique, resulting in poor chromatography of soil samples, shorter lifetimes for GC columns and significant breakdown of endrin and DDT during analysis. The new protocol requires that GPC be used for all soil samples. It is now required that all extracts be subjected to an alumina column cleanup, but some lots of alumina remove pesticides added as matrix spikes in QC samples. This is strong evidence that the requirement for alumina cleanup causes lower recoveries of analytes from samples due to the high reactivity of the adsorbent. The new protocol allows the option of open alumina or Florisil columns or the option of prepacked Diol bonded silica cartridge columns (Analytichem or equivalent) for extract cleanup.

The present requirement for GC performance checks that measure analyte breakdown and detector linearity cannot be met by many laboratories using the present acceptance criteria. The breakdown of endrin and DDT on column can be extensive, especially when the injector and/or the column are contaminated [8]. The requirement in the new protocol for GPC cleanup of all soil samples should reduce the breakdown problem significantly.

The present protocol requires that a one-point calibration factor be determined for each analyte and that a periodic linearity check of the instrument be run using three concentrations of endrin, DDT, aldrin and DBC to give chromatograms with peaks of 20, 50 and 100 percent of full scale. In order to achieve an acceptable linearity check, the calibration curve for each of the four compounds must pass through the origin, which is a difficult requirement to satisfy. The new protocol requires that a linear calibration curve initially be established for each analyte. The laboratory must also confirm that the response of a low concentration of two BHC isomers, medium concentrations of endrin and aldrin and a high concentration of DDT are within acceptance windows every twelve hours. These requirements will ensure that the detector response is always linear for each analyte and should be easier to satisfy than the present requirement.

There are also problems with the number of calibration runs required in present protocol. Originally a complete calibration for all analytes was required every 24 hours. This meant that for eight of every 24 hours each laboratory had to collect calibration data. In 1985 this requirement was changed to a 72 hour calibration period, but even so, an excessive amount of time is spent by each laboratory calibrating standards without significantly increasing data quality. The new protocol allows an initial calibration to be used until the laboratory is unable to meet the acceptance criteria for the periodic (12 hour) performance evaluation described in the QA/QC section below.

DESCRITPION OF THE NEW PESTICIDE/PCB PROTOCOL

Scope

The proposed protocol is suitable for analysis of the 17 single component pesticides and nine multicomponent pesticide/PCBs in soil and water above the detection limits given in Table 1. The protocol is presently being validated in a single laboratory study by S-CUBED. Single laboratory validation will be followed by a limited multilaboratory study and a full collaborative study. Elements of the procedure were taken from the present protocol [4], EPA Method 608 [5], Method 8080 [6] and the pesticide method of EPA Region IV [7]. This protocol should give data of known quality and be suitable for use in a production mode by CLP laboratories in support of the Superfund program.

Extraction of Water Samples

Before extraction, the pH of the sample is measured and adjusted to between pH 5 and pH 9 if required. Both the initial and final pH are reported with the sample data. After pH adjustment, a 1-L water sample is extracted with methylene chloride using either a separatory funnel or a continuous liquid-liquid extractor. A continuous extractor is required if emulsions are formed using a separatory funnel. The methylene chloride extract is dried with sodium sulfate, reduced in volume with a Kuderna-Danish (K-D) apparatus, and exchanged to hexane unless GPC cleanup is to be used.

Extraction of Soil/Sediment Samples

Before the extraction of soil or sediment samples, any standing water is decanted from the sample. The pH and the percent moisture are then determined and reported with the sample data, but no adjustment of sample pH is used in the soil/sediment procedure. A 30-gram sample of soil is mixed with 60-grams of dry sodium sulfate and extracted with methylene chloride/acetone (1:1) three times using an ultrasonic extractor probe in the pulsed mode. The use of the ultrasonic extractor is an attractive alternative to traditional soxhlet extraction because of the increase in sample throughput using the sonicator method without sacrificing extraction efficiency. It has been shown that there is no significant difference in the recovery of organochlorine pesticides/PCB's from soils using the two techniques [9].

Following extraction of the soil sample with three portions of methylene chloride/acetone, the organic fraction is dried with sodium sulfate and reduced to a volume of 10-mL with a K-D apparatus.

Extract Cleanup

GPC cleanup is mandatory for all soil/sediment extracts and may be used for water samples at the option of the laboratory. GPC removes many higher molecular weight compounds which would otherwise contaminate the GC causing analyte breakdown or cause interfering peaks in sample chromatograms.

GPC cleanup is accomplished with a calibrated column packed with SX-3 Bio-Beads using methylene chloride as a solvent. Higher molecular weight materials elute first and are discarded. The fraction containing the pesticides and PCB's is collected and the volume of methylene chloride is reduced in a K-D apparatus. After evaporation of methylene chloride, a portion of hexane is added and the volume reduced again in order to exchange the solvent to hexane before column cleanup. The GPC system performance must be validated every 30 days using the procedure described in the QA/QC section of this paper.

Following GPC cleanup, polar interferents are removed from the extract using an absorption column. As discussed above, problems have been observed with the alumina cleanup procedure required in the present pesticide/PCB method. Because some analytes do not elute through some lots of alumina, the new method also allows Florisil or cartridge cleanup. Florisil chromatography had been used for pesticide analysis since the 1950's [10, 11] and is used in several methods validated for use under the auspices of the EPA [12, 13]. Because of the large volume of analyses required of CLP laboratories, an alternative cleanup procedure using prepacked Diol bonded silica columns (Analytichem, Harbor City, Califorina) was validated. Using this cleanup technique, up to ten samples can be prepared simultaneously using cartridges with prepacked adsorption beds held in place with stainless steel frits. The columns are free of contaminants and greatly increase the throughput of the method.

Samples containing sulfur must undergo a further cleanup using one of three options. The option that appears to be the most reliable requires adding one to three drops of mercury to each extract in a clean vial. The vial is then agitated and the organic solvent removed for GC analysis. If the laboratory does not wish to use mercury, copper can be substituted. The third option is to remove the sulfur from the sample using a tetrabutylammonium sulfate solution which is prepared by the laboratory from tetrabutylammonium hydrogen sulfate and sodium sulfide in water.

GC Analysis

The present CLP protocol for pesticides/PCB's requires dual GC analysis, the protocol being validated allows both packed column or capillary column analysis. The primary packed column is an 8-foot x 4-mm ID glass column packed with 1.5 percent SP2250/ 1.95 percent SP2401 (or equivalent) on 100/120 mesh Supelcoport. The secondary column is 10-foot x 4-mm ID glass column packed with 5 percent SP2100 (or equivalent) on 100/120 mesh Supelcoport. Packed column analysis is accomplished using isothermal operation of both columns at 205°C. Capillary GC analysis of the HSL pesticides/PCB's is accomplished with a DB-608 or SPB-608 (or equivalent) 30 meter x 0.53 mm "megabore" capillary column as a primary column and with a 30 meter x 0.53 mm DB-5 (or equivalent) secondary column. Capillary analysis is accomplished using temperature programming from an initial temperature of 50° to a maximum of 200°C with a ramp of 8°/min. The temperature program for capillary analysis must include an initial temperature hold of three minutes if a dual column oven is used, ensuring injection of both columns prior to initiation of the temperature ramp.

An initial calibration is required for each column used for analysis. This is used to establish the retention times relative to the internal standard dibutylchlorendate and a response factor for each analyte. In order to assure that each component is adequately resolved during the calibration runs, the laboratory must prepare the single component standards in the two separate individual mixes given in Table 2. The medium concentration used is 10 times the low concentration given in Table 2. The high concentration is 100 times the low concentration. Calibration data must be collected from on scale chromatograms.

TABLE 2--Calibration Mixtures for Single Component Pesticides

Individual Standard Mix A	Low Concentration ng/mL	Individual Standard Mix B	Low Concentration ng/mL
alpha-BHC	2.5	beta-BHC	2.5
heptachlor	2.5	delta-BHC	2.5
gamma-BHC	2.5	aldrin	2.5
endosulfan I	2.5	heptachlor epoxide	2.5
dieldrin	5.0	p,p'-DDE	5.0
endrin	5.0	endosulfan sulfate	5.0
p,p'-DDD	5.0	endrin aldehyde	5.0
p,p'-DDT	5.0	endrin ketone	5.0
methoxychlor	25	endosulfan II	5.0
dibutylchlorendate	25	dibutylchlorendate	25
		hexabromobenzene	2.5

The identification of all analytes is based on their retention time relative to dibutylchlorendate which is added to all samples, blanks and standards injected. All identified analytes must fall in a window of ±1.5 percent of the mean relative retention time of standards injected during the initial calibration. If dibutylchlorendate cannot be observed in a sample chromatogram due to matrix interference, identification is based on an absolute retention time window of ±2 percent of the mean retention time of the standards established during the initial calibration.

All sample data must be collected in a run sequence which includes an initial calibration sequence, method blanks, matrix spike and matrix spike duplicate analyses, instrument blanks and periodic evaluation standards. The specific requirements for the run sequence are discussed below in QA/QC requirement section.

The method also specifies data acceptance criteria for sample chromatograms. These include the requirements that the baseline of an acceptable chromatogram return to below 50 percent of full scale before the elution of alpha-BHC and to below 25 percent of full scale before the elution of dibutylchlorendate. In addition, all reported pesticides must be within the linear range determined during the initial calibration. If any analyte is detected at a concentration above the linear range, the sample must be diluted and reinjected.

Sample chromatograms must be presented with all major peaks on scale. If the laboratory has an electronic data system capable of replotting data, the on scale chromatograms may be replotted from data collected during the initial analysis. If chromatograms cannot be replotted, the extract must be diluted and reanalyzed in order to being the peaks on scale. Sample chromatograms with interfering peaks or high base lines that prevent quantitation of analytes must be reanalyzed after additional cleanup and/or dilution. Whenever chromatograms must be replotted or samples diluted to bring peaks on scale, the initial analytical run must also be submitted to the EPA.

Identification and Quantitation of Analytes

Prior to analyzing samples each single component pesticide must be run at three different concentrations in order to establish the retention time relative to the internal standard and to establish the response factor for each compound in the linear range of the detector. The low point for the calibration curve corresponds to the contract required detection limit (CRDL) given in Table 1. The midpoint of the calibration curve is 10 times the CRDL and the high point of the calibration curve must be at least 100 times the CRDL. Multicomponent pesticide and PCB standards are run at a single concentration and must be run individually except for Aroclor 1260 and Aroclor 1016, which are run as a standard mixture.

The single component analytes are identified based on their retention time relative to the internal standard (RRT). On packed columns they must be within 1.5 percent of the mean RRT established during initial calibration and on capillary columns they must be within 0.5 percent. The concentration of single component pesticides are calculated using the equations given below:

$$\text{Concentration in water } \mu g/L \ = \ \frac{(A_x)(V_t)}{(RF)(V_i)(V_x)} \tag{1}$$

where

A_x = Detector response for the parameter to be measured.

RF = Response factor for the external standard (detector response/ng).

V_t = Volume of total extract (μL) (take into account any dilution).

V_i = Volume of extract injected (μL).

V_x = Volume of water extracted (mL).

$$\text{Concentration in soil } ug/kg \ = \ \frac{(A_x)(V_t)}{(RF)(V_i)(W_s)(D)} \tag{2}$$

(Dry weight basis)

where

A_x, RF, V_i = same as given above in Equation 1.

$$D = \frac{100 - percent\ moisture}{100}$$

W_s = Weight of sample extracted (gram)

The identification and quantitation of multicomponent analytes is based on both visual pattern recognition and on measurement of three major peaks of the analyte on the packed columns. Analyte concentration is calculated using equations 1 and 2 where A_x is based on the sum of three analyte peaks (heights or areas) and RF on the sum of the same three peaks in the standard. Table 3 gives the relative retention time for three major peaks of each multicomponent analyte on both packed columns at 205°C with 30 mL/min P-5 carrier gas flow. The choice of peaks used for the quantitation of any multicomponent analyte may be complicated by its environmental degradation and by the presence of coeluting analytes or matrix interference.

TABLE 3--RRTs VERSUS DIBUTYLCHLORENDATE FOR SELECTED MAJOR PEAKS IN MULTICOMPONENT ANALYTES

Analyte	RRT on 1.5 Percent SP-2250 1.95 Percent SP-2401	RRT on SP-2100
Aroclor 1016	0.08, 0.10, 0.13	0.09, 0.11, 0.14
Aroclor 1221	0.07, 0.08, 0.10	0.06, 0.08, 0.09
Aroclor 1232	0.08, 0.13, 0.23	0.09, 0.14, 0.24
Aroclor 1242	0.10, 0.13, 0.23	0.11, 0.14, 0.24
Aroclor 1248	0.13, 0.16, 0.23	0.14, 0.17, 0.24
Aroclor 1254	0.23, 0.40, 0.51	0.23, 0.39, 0.45
Aroclor 1260	0.57, 0.88, 1.1	0.46, 0.53, 0.83
Chlordane	0.13, 0.18, 0.26	0.16, 0.26, 0.29
Toxaphene	0.35, 0.51, 0.62	0.32, 0.40, 0.52

QA/QC REQUIREMENTS

The implementation of a rigid QA/QC requirement is an important part of any method. The QA/QC requirements of this protocol are necessary to assure that the data collected are of known quality, to allow interlaboratory comparison of data to facilitate including all data in a national data base of environmental contamination, and to ensure that all data collected are legally defensible. The QA/QC program of the protocol presently being validated has two aspects, requirements for instrument performance and requirements for method performance. Each laboratory must document satisfactory completion of these requirements on standard forms.

Instrument QA/QC

Prior to the analysis of any blanks or samples, the laboratory must run a three-point initial calibration for all single component analytes and a single point determination of all multiple component analytes. For each single component analyte the mean deviation for the three absolute retention times must be less than 0.3 percent of the average of the retention times and the relative standard deviation of the response factors must be less than or equal to 20 percent of the average of the three determinations. Each peak in the chromatograms of standard mixtures A and B (Table 2) must be at least 75 percent resolved.

Analyte retention time, calibration, resolution and stability must be monitored every 12 hours by injection of the six compounds listed in Table 4. For each analyte, the retention time relative to the internal standard dibutylchlorendate must be within 1.5 percent of the average of the calibration standard and the response factor must be within 20 percent of the calibration average. Endrin and DDT are included as analytes that have the potential to degrade on column [8], their combined breakdown must be less than 20 percent. Beta-BHC and gamma-BHC are included as resolution checks. The height of the valley between the two BHC peaks must be no more than 10 percent of the height of the taller BHC peak on the SP2250/SP2401 column, not greater than 25 percent on the SP2100 column nor more than 10 percent on either of the capillary columns. At least one acceptable evaluation standard must be run every 12 hours. If an acceptable evaluation standard cannot be run, the chromatographic system is considered out of control and appropriate corrections must be made. After the corrections are made the initial calibration must be rerun before any sample data are collected. Any data that are not bracketed with acceptable evaluation runs are not acceptable, and those samples must be reinjected.

TABLE 4--PERFORMANCE EVALUATION MIXTURE

Pesticide	Concentration
Beta-BHC	2.5 ng/mL
Gamma-BHC	2.5 ng/mL
Aldrin	25 ng/mL
4,4'-DDT	500 ng/mL
Endrin	50 ng/mL
Dibutylchlorendate	250 ng/mL

All multicomponent analytes must be run periodically during a run sequence to confirm that the initial calibration of those compounds is still valid. Toxaphene, chlordane, and Aroclors 1016, 1242, 1248, 1254 and 1260, must be run every week. Aroclors 1221 and 1232 must be run monthly. If the combined response factor of any of the multicomponent analytes is more than 20 percent different from the initial calibration value, the GC system is considered out of control and appropriate corrective action must be taken. After the corrections are made, the initial calibration must then be rerun and all samples with detectable levels of PCB's run since the last acceptable set of standards must be rerun.

The cleanup of all samples requires adsorption chromatography with Florisil or alumina columns or with Diol bonded silica cartridges. Every batch of adsorbent must be tested by eluting 2,4,5-trichlorophenol and standard mixture A (Table 2) through a column as if it were an extract. Adsorption column performance is acceptable if all pesticides are recovered at 80-110 percent and if no trichlorophenol is detected.

GPC calibration must be confirmed every 30 days. Two solutions are required, one containing the matrix spiking solution listed in Table 5 and the second containing a mixture of Aroclor

1016 and Aroclor 1260. After elution through the GPC, the recovery of all analytes must be in the 80-110 percent range.

TABLE 5--MATRIX SPIKING SOLUTION

Pesticide	µg/1.0 mL
Gamma-BHC	0.5
Heptachlor	0.5
Aldrin	0.5
Dieldrin	1.0
Endrin	1.0
4,4'-DDT	1.0

Sample QA/QC

Blanks: A method blank must be analyzed for every 20 samples extracted by each of the three extraction procedures (continuous liquid-liquid, separatory funnel or sonic extraction) used in a case (defined as a group of samples from a given site taken during the same time period). An acceptable method blank has no analyte present at a level greater than the CRDL (Table 1). If an unacceptable method blank is run, all samples analyzed since the last acceptable method blank must be reextracted and reanalyzed.

Instrument blanks are required for both the GC and the GPC. An instrument blank is run by analyzing an injection of solvent. An acceptable instrument blank must have no analyte present at greater than one-half the CRDL (Table 1). If an unacceptable instrument blank is run, corrective action must be taken and all samples run since the last acceptable instrument blank must be reanalyzed.

Matrix Spike/Matrix Spike Duplicate Analysis: A matrix spike/matrix spike duplicate (MS/MSD) analysis must be performed for every 20 samples prepared by each extraction procedure used in a case or at least once for each matrix type in a case. The MS/MSD pair is a replicated sample prepared in the laboratory by adding matrix spike solution (Table 5) to duplicate aliquots of a sample prior to extraction. The MS/MSD pair are extracted, cleaned up and analyzed according to the CLP protocol described above. The recoveries of the individual compounds spiked are corrected for the values detected in the unspiked sample and reported to the EPA as a check on method performance. The laboratory must also report the relative percent difference between the matrix spike and the matrix spike duplicate in order to provide the EPA with information on the precision of the method. Recommended performance based on QC limits for both the relative percent difference and the percent recovery of the matrix spike compounds are given to the laboratories.

Surrogate Analysis: The surrogate hexabromobenzene (HBB) must be added to all samples and method blanks prior to extraction in order to monitor method performance and sample chromatography. HBB was chosen because it chromatographs near the middle of the analytical run, it has a different retention time from commonly observed matrix interferences and HSL pesticides, and it does not undergo   ester   hydrolysis   unlike   the   present   surrogate, dibutylchlorendate.

METHODOLOGY OF LABORATORY VALIDATION OF THE PROTOCOL

Validation of the CLP pesticide/PCB protocol is being conducted according to EPA guidelines presented in "Validation of the Testing/Measurement Methods" [14] and those of the Association of Official Analytical Chemists (AOAC) [15]. Validation will proceed in three steps, a single laboratory study, a limited multilaboratory study and a (proposed) full multilab study.

The single laboratory study will be completed as a first step in the process. During this single laboratory study, the ruggedness (sensitivity of the method to small changes in critical method variables) will be investigated according to the scheme described by the AOAC. Once the method has been tested for ruggedness, a number of spiked environmental samples will be analyzed and the precision, the systematic error (bias), upper and lower detection limits, and sensitivity (ability to detect a 10 percent change in analyte concentration) of the method will be determined. At the conclusion of the single laboratory portion of the study, the protocol will be rewritten to include any improvements in the method. The revised protocol will be sent out for review within the EPA and by other laboratories.

The limited multilaboratory validation of the method will then be conducted. Participating laboratories will analyze the six groups of spiked water and soil samples listed in Table 6, and report the data to S-CUBED for statistical evaluation of method precision, bias, and sensitivity. In addition, the laboratories will make comments about difficulties in the operation of the method.

TABLE 6 --SAMPLES TO BE ANALYZED IN LIMITED MULTILAB VALIDATION OF THE METHOD

Identification	Description
A	Low concentration - single component pesticides
B	Medium concentration - single component pesticides
C	High concentration - single component pesticides
D	Medium concentration - multicomponent pesticides
E	Low sensitivity - low concentration +10 percent
F	Medium sensitivity - medium concentration +10 percent

If the data collected during the multilaboratory validation are of sufficiently high quality and the method can be used routinely, the method may be tested by CLP laboratories in a full multilaboratory validation. These laboratories will analyze samples in order to generate Youden pair data for analytes selected by the EPA as a last step in the validation of the method.

CONCLUSION

Although the current pesticide/PCB method has been used successfully by CLP laboratories to produce high quality data that has been used in remedial actions and for enforcement of the liability provisions of CERCLA, the method has never been properly validated. The development of the new method for pesticides/PCB's described here is the first step in the adoption of fully validated protocols for use in all CLP analyses. The development of validated methods is just part of the EPA's ongoing effort to improve the quality of data provided by the CLP program as part of its mandate to remove hazardous pollutants from the environment.

Acknowledgement

This method has been developed by a large group of people who should appear as coauthors, they include: Phil Ryan, Ken Baughman, Joan Fisk, William Loy, Lynn Williams and Mike Scott. The work was conducted under the sponsorship of the U.S. EPA through the Superfund Program and the Environmental Monitoring and Systems Laboratory - Las Vegas (contract number 68-03-1958).

This method has not been subjected to policy review by the U.S. EPA and does not necessarily reflect the views of the Agency. Mention of specific trade names or commercial products is for identification purposes only and does not consititue endorsement of nor recommendation for use.

REFERENCES

[1]   Fisk, J.F., "Semivolatile Organic Analytical Methods - General Description and Quality Control Considerations", Quality Control in Remedial Site Investigation: Hazardous and Industrial Solid Waste Testing, Fifth Volume, ASTM STP 925, American Society of Testing Materials, this publication.

[2]   Fisk, J.F., "Volatile Organic Analytical Methods General Description and Quality Control Considerations", Quality Control in Remedial Site Investigation: Hazardous and Industrial Solid Waste Testing, Fifth Volume, ASTM STP 925, American Society of Testing Materials, this publication.

[3]   Fowler, J.W., "Inorganic Analytical Methods - General Description and Quality Control Consdierations, Quality Control in Remedial Site Investigation: Hazardous and Industrial Solid Waste Testing, Fifth Volume, ASTM STP 925, American Society of Testing Materials, this publication.

[4]   U.S. EPA, "Statement of Work for Organics Analysis - Multi-Media Multi-Concentration, Revision 9/85, U.S. EPA, Laboratory Program, Office of Emergency and Remedial Response, Washington.

[5]   Pressley, T.A., and Longbottom, J.E., "The Determination of Organochlorine Pesticides and PCB's in Industrial and Municipal Wastewater, Method 608", Report EPA-600/4-82-057, Environmental Monitoring and Support Laboratory, Cincinnati, OH, 1982.

[6]   Anon., "Method 8080-Organochlorine Pesticides and PCB's", Manual SW-846, Office of Solid Waste and Emergency Response, U.S. EPA, Washington, D.C., 1982.

[7]   Loy, E.W., "Section 10 - Performance Quality Control and Operations - Organic Analysis", Analytical Support Branch Operation and Quality Control Manual, U.S. EPA Region 4, Athens, GA, 1985.

[8]   Bellar, T.A., Stemmer, P., and Lichtenberg, J.J., "Evaluation of Capillary Systems for the Analysis of Environmental Extracts", EPA Report No. 3555C, The Office of Drinking Water, U.S. EPA, Washington, D.C., 1983.

[9]   Warner, J.S., Landes, M.C., Slivon, L.E., "Development of a Solvent Extraction Method for Determining Semivolatile Organic Compounds in Solid Wastes", in Hazardous and Industrial Solid Waste Testing: Second Symposium, ASTM STP 805, American Society of Testing Materials, Philadelphia, 1983, pp. 203-213.

[10]   Ordas, E.P., Smith, V., and Meyer, C.F., "Spectrophotometric Determination of Heptachlor and Technical Chlordane on Food and Forage Crops", Journal of Agriculture and Food Chemistry, Vol. 4, 1956, pp. 444-451.

[11]   Mills, P.A., "Determination and Semiquantitative Estimation of Chlorinated Organic Pesticide Residues in Foods by Paper Chromatography", Journal of the Association of Official Analytical Chemists, Vol. 42, 1959, pp. 734-740.

[12]   Mills, P.A., Onley, J.H., and Gaither, R.A., "Rapid Method for Chlorinated Pesticide Residues in Nonfatty Foods", Journal of the Association of Official Analytical Chemists, Vol. 46, 1963, pp. 186-191.

[13]   McMahon, B., and Burke, J.A., "Analytical Behavior Data for Chemicals Determined Using AOAC Multiresidue Methodology for Pesticide Residues in Foods", Journal of the Association of Official Analytical Chemists, Vol. 61, No. 3, 1978, pp. 640-652.

[14]   Williams, L.R., "Validation of Testing/Measurement Methods", EPA 600/X-83-060, U.S. EPA, Las Vegas, NV, 1983.

[15]   Youden, W.J. and Steiner, E.H., Statistical Manual for the AOAC, Association of Official Analytical Chemists, Arlington, VA, 1975.

Author Index

Subject Index